Decision Aid Framework Development to Inform the Application of Federal Assistance

Improving Sharing of Criminal History Record Information for Personnel Vetting

DAVID LUCKEY, BRADLEY M. KNOPP, AMANDA WICKER, DIANA Y. MYERS, WILL SHUMATE, CHRIS DICTUS, BRANDON DE BRUHL, DAVID STEBBINS

Prepared for the Defense Counterintelligence and Security Agency
Approved for public release; distribution is unlimited

For more information on this publication, visit **www.rand.org/t/RRA891-1**.

About RAND

The RAND Corporation is a research organization that develops solutions to public policy challenges to help make communities throughout the world safer and more secure, healthier and more prosperous. RAND is nonprofit, nonpartisan, and committed to the public interest. To learn more about RAND, visit www.rand.org.

Research Integrity

Our mission to help improve policy and decisionmaking through research and analysis is enabled through our core values of quality and objectivity and our unwavering commitment to the highest level of integrity and ethical behavior. To help ensure our research and analysis are rigorous, objective, and nonpartisan, we subject our research publications to a robust and exacting quality-assurance process; avoid both the appearance and reality of financial and other conflicts of interest through staff training, project screening, and a policy of mandatory disclosure; and pursue transparency in our research engagements through our commitment to the open publication of our research findings and recommendations, disclosure of the source of funding of published research, and policies to ensure intellectual independence. For more information, visit www.rand.org/about/research-integrity.

RAND's publications do not necessarily reflect the opinions of its research clients and sponsors.

Published by the RAND Corporation, Santa Monica, Calif.

Library of Congress Cataloging-in-Publication Data is available for this publication.

ISBN: *978-1-9774-0938-6*

About This Report

The RAND Corporation's National Defense Research Institute (NDRI) conducted research and analysis to develop a decision aid framework (DAF) to assist the U.S. Department of Defense's Defense Counterintelligence and Security Agency (DCSA). We identified critical factors that enable or inhibit the provision of criminal history record information (CHRI) to DCSA from state, local, tribal, and territorial (SLTT) law enforcement agencies (LEAs) whose CHRI contributions are needed for vetting individuals as part of the background investigation process. In creating the DAF, we developed a process that would allow DCSA, when assessing which localities might benefit from targeted federal assistance, to prioritize SLTT LEA organizations according to DCSA goals.

The research reported here was completed in April 2022 and underwent security review with the sponsor and the Defense Office of Prepublication and Security Review before public release.

RAND National Security Research Division

This research was sponsored by the Defense Counterintelligence and Security Agency and conducted within the Forces and Resources Policy Center of the RAND National Security Research Division (NSRD), which operates the National Defense Research Institute (NDRI), a federally funded research and development center sponsored by the Office of the Secretary of Defense, the Joint Staff, the Unified Combatant Commands, the Navy, the Marine Corps, the defense agencies, and the defense intelligence enterprise. For more information on the Forces and Resources Policy Center, see www.rand.org/nsrd/frp or contact the Director (contact information is provided on the web page). Comments or questions on this report should be addressed to the project leaders, David Luckey at dluckey@rand.org and Bradley M. Knopp at bknopp@rand.org.

Acknowledgments

The authors thank the director of DCSA, William K. Lietzau, and staff, especially Dan Leary, for their interest in this project; Daniel Ginsberg, Rich Girven, and Charles Sowell for their reviews of the report; Molly McIntosh for providing the opportunity to undertake this research; and others for providing advice and expertise throughout the project.

We also thank researchers at the Integrated Justice Information Systems Institute and DCSA Federal Investigative Records Enterprise, Field Ops, Law Enforcement Liaison Office, and headquarters staff for their valuable insights that have influenced this report.

Although we are grateful for the assistance by those provided above, please attribute any errors or omissions in this report solely to the authors.

Summary

The RAND Corporation's National Defense Research Institute conducted research and analysis to develop a decision aid framework (DAF) to assist the U.S. Department of Defense's Defense Counterintelligence and Security Agency (DCSA). We identified critical factors that enable or inhibit the provision of criminal history record information (CHRI) to DCSA from state, local, tribal, and territorial (SLTT) law enforcement agencies (LEAs) whose CHRI contributions are needed for vetting individuals as part of the background investigation (BI) process. In creating the DAF, we developed a process that would allow DCSA, when assessing which localities might benefit from targeted federal assistance, to prioritize SLTT LEA organizations according to DCSA goals.

Approach

SLTT LEA CHRI is critical to DCSA's BI process, yet SLTT LEAs encounter a variety of barriers to CHRI sharing, which in turn hinders the vetting process. To assess which localities might benefit from targeted federal assistance, we developed a DAF that identifies and prioritizes SLTT LEA organizations based on the kind of assistance needed.

We completed our work in three main tasks. The first task included a survey of the statutory and regulatory authorities that govern the current process. A main effort of this task was to identify impediments that currently constrain DCSA collaboration with SLTT partners. For the second task, we analyzed data provided by DCSA to identify obstacles to efficient information-sharing and the SLTT organizations whose contributions are most important for vetting and processing BIs. In the final task, we employed visualization tools and advanced analytic techniques to display the analysis to help DCSA leadership make data-driven decisions regarding where intensified cooperation and application of federal resources could mitigate CHRI collection issues and thereby improve the efficiency of the vetting process.

Findings and Recommendations

The purpose of our research was to develop a tool for use by DCSA. In addition, we developed the following findings and recommendations:

- **DCSA's CHRI collection is complicated by the myriad SLTT laws and policies regarding CHRI collection, maintenance, and sharing.** In particular, different jurisdictions have different laws for expunging and sealing records, reporting arrests and dispositions, collecting fingerprints, and automatically clearing records.
  - Recommendations:
    - Improve data collection and enable cooperation by providing LEAs with access to federal funding and assistance that could be used to modernize criminal history data collection, management, and sharing capabilities. New data collection acquisition and management capability would directly support DCSA efforts to streamline and improve and possibly accelerate personnel vetting processes.
    - Ensure all DCSA investigators are trained on local CHRI laws and policies so that they can address specific SLTT concerns about sharing CHRI.

- **SLTT information exchange between SLTT LEAs and DCSA is critical to DCSA's vetting process; unfortunately, many SLTT LEAs lack the personnel, knowledge, technology, or funding to comply with their obligation to report CHRI in support of federal vetting.** Outdated and outmoded reporting systems frequently slow or obstruct DCSA's CHRI collection, and manual CHRI sharing could impose additional burdens on both LEA employees' and DCSA investigators' time.
  - Recommendations:
    - Improve cooperation by providing LEAs with access to federal funding and assistance that could be used to modernize criminal history data collection, management, and sharing capabilities.

    - Provide training, education, and direct assistance to SLTT LEAs for the purpose of streamlining access to historical and future CHRI data.[1]

- **The CHRI collection challenges that SLTTs face might be a product of SLTT laws and policies, the administrative burden the employees face in sharing CHRI, systemic barriers, government interaction through participation in federal programs, and structural conditions.** By focusing on the potential causes of insufficient CHRI sharing, DCSA would be better positioned to mitigate those specific collection issues. Additional information regarding the interactions of SLTTs with DCSA investigators (staff and contractors) was not available to the researchers, but would almost certainly improve the utility of the DAF.
  - Recommendations:
    - Use the DAF to drive decisions on where intensified cooperation and application of federal resources could best mitigate CHRI collection issues and ultimately improve the efficiency and effectiveness of DCSA CHRI collection.
    - Collect data on SLTT-DCSA interactions for all DCSA investigators and update the data when challenges are resolved or new information becomes available. Ensure data quality and accuracy by implementing standards for entry.

---

[1] For more detailed recommendations on SLTT LEA education and training, see Ligor, Douglas C., Shawn Bushway, Maria McCollester, Richard H. Donohue, Devon Hill, Marylou Gilbert, Heather Gomez-Bendaña, Daniel Kim, Annie Brothers, Melissa Baumann, Barbara Bicksler, Rick Penn-Kraus, and Stephanie Walsh, *Criminal History Record Information Sharing with the Defense Counterintelligence and Security Agency: Education and Training Materials for State, Local, Tribal, and Territorial Partners*, Santa Monica, Calif.: RAND Corporation, RR-A846-1, 2022.

Contents

Figures and Tables

Figures

Tables

CHAPTER ONE

Introduction

Purpose

State, local, tribal, and territorial (SLTT) organizations provide criminal history record information (CHRI) for background investigations (BIs): data that is required for vetting current and prospective government employees, military service members, and contractor personnel.[1] There are more than 18,000 law enforcement agencies (LEAs) in the United States, each housing various amounts of CHRI.[2] These data are vital to the proper and full vetting of persons under suitability and fitness standards, including standards for determining eligibility for access to national security classified information, assignment to positions with sensitive duties, and access to government facilities and systems.[3] Using statutory authority, delegations, and directions, the U.S. Department of Defense's (DoD's) Defense Counterintelligence and Security Agency (DCSA) conducts approximately 95 percent of all BIs on behalf of U.S. government departments and agencies.[4]

---

[1] National Consortium for Justice Information and Statistics (SEARCH), *Challenges and Promising Practices for State Criminal History Repositories: Report to the Performance Accountability Council (PAC) Program Management Office (PMO)*, March 24, 2020a.

[2] President's Task Force on 21st Century Policing, *Final Report of the President's Task Force on 21st Century Policing*, Washington, D.C.: U.S. Department of Justice, No. 248928, May 2015.

[3] Homeland Security Presidential Directive 12, *Policy for a Common Identification Standard for Federal Employees and Contractors*, Washington, D.C.: U.S. Department of Homeland Security, August 12, 2004.

[4] See Executive Order 13869, *Transferring Responsibility for Background Investigations to the Department of Defense*, Washington, D.C.: The White House, April 24, 2019; and DCSA, *DCSA Background Investigation Overview*, September 22, 2020c.

The DCSA Law Enforcement Liaison Office (LELO) is working to address persistent challenges associated with federal background investigators' access to SLTT CHRI, both for its own purposes and on behalf of the approximately 20 other federal agencies that conduct vetting investigations. Information exchange between SLTT LEAs and DCSA is critical to DCSA's vetting process. Unfortunately, many SLTT LEAs lack the personnel, knowledge, technology, or funding to comply with their obligation to report CHRI in support of federal vetting.[5] DCSA, at the outset of our study, explained that it had federal money to invest in assisting SLTT organizations to modernize criminal history data collection, management, and sharing capabilities and improve cooperation with DCSA.[6] These new data acquisition and management capabilities would directly support DCSA efforts to streamline, improve, and possibly accelerate personnel vetting processing. Although DCSA was provided funding for this purpose, it lacked both a mechanism and the authority to disperse such funds. In addition, it did not possess a systemic approach for doing so. As a result, DCSA sought research and analytic support from the RAND Corporation to develop a framework and systemic approach to improve funding disbursement. This report provides both.

Pursuant to Section 1625 of the National Defense Authorization Act for Fiscal Year 2020, the director of DCSA is to reduce the time and cost of SLTT CHRI for BIs required for vetting current and prospective government employees, military service members, and contractor personnel.[7] The activities carried out in pursuit of this goal include training, education, and direct assistance to SLTT LEAs for the purpose of streamlining access to historical CHRI data. To assist DCSA in executing this remit, we identify critical factors that enable or inhibit the provision of CHRI to DCSA from

[5] For more on the federal law that requires CHRI sharing with federal government agencies for background investigations, see U.S. Code, Title 5, Section 9101, Access to Criminal History Records for National Security and Other Purposes, 2011.

[6] DCSA, *Report on the Defense Counterintelligence and Security Agency's Personnel Vetting Mission: In Response to H.R. 2500, the U.S. House of Representatives-Passed Fiscal Year 2020 NDAA*, June 2020b.

[7] Public Law 116-92, National Defense Authorization Act for Fiscal Year 2020, December 20, 2019.

SLTT LEAs whose CHRI contributions are needed for vetting individuals as part of the BI process and who might benefit from federal assistance. The accompanying decision aid framework (DAF) is intended to help DCSA to prioritize SLTT LEA organizations according to DCSA goals and priorities.

Background

DCSA's criminal history record checks for national security purposes are more extensive than other employment-related checks because of the sensitivity of the information handled. Because the investigation process relies heavily on centralized, nationwide CHRI repositories, such as those managed by the Federal Bureau of Investigation (FBI), the National Crime Information Center (NCIC), and the National Law Enforcement Telecommunications System (Nlets), the efficiency and accuracy of an investigation relies on the completeness and availability of CHRI data inputted into the system.[8] Throughout an investigation, completeness and accuracy of CHRI data is important because it reduces the bureaucratic and administrative burdens of DCSA officers who must contact local LEAs when needed data is missing. As shown in Figure 1.1, if background investigators and adjudicators could get sufficient information from centralized databases (e.g., Nlets) during the automated record check phase, the amount of fieldwork needed to obtain missing or incomplete information would be reduced.[9] This figure shows a simplified process. The more information gathered in the automated portion of the process, the quicker the case can move to adjudication.

---

[8] Nlets is a private, nonprofit organization that provides law enforcement (LE) and criminal justice agencies with various criminal history information, including data on driver's license and Interpol reporting. DCSA supplements its Interstate Identification Index (III) and NCIC checks with Nlets information. See Shawn Bushway, Douglas Ligor, and Stephanie Walsh, *Process Review: Overview, Initial Observations & Potential Responses for DCSA's State and Local CHRI Acquisition Process*, Santa Monica, Calif.: RAND Corporation, forthcoming.

[9] DCSA, 2020c.

FIGURE 1.1
Simplified National Security Background Investigation Process Map

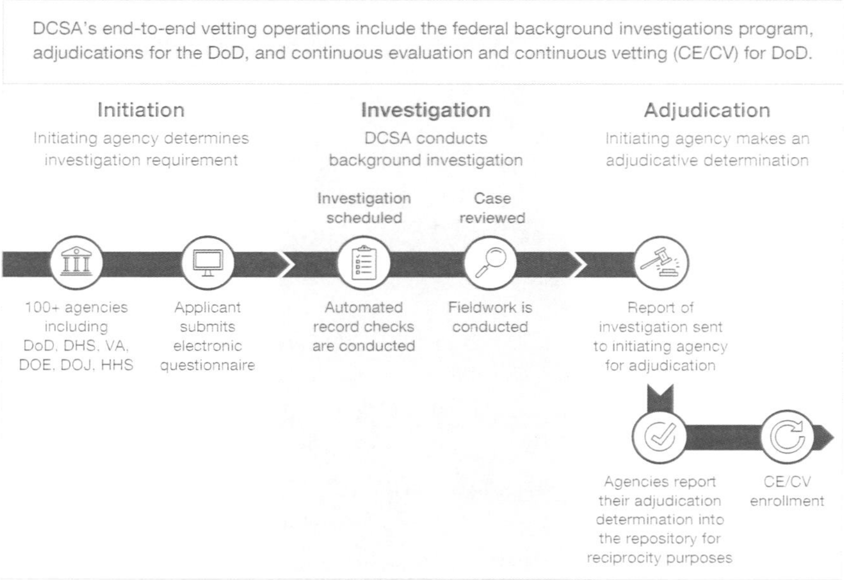

SOURCE: Adapted from DCSA, 2020c.
NOTE: DHS = U.S. Department of Homeland Security, DOE = U.S. Department of Education, DOJ = U.S. Department of Justice, HHS = U.S. Department of Health and Human Services, VA = U.S. Department of Veterans Affairs.

Method

To assess what SLTT organizations must do to support DCSA's BI process and what resources are required to accomplish this mission, we developed a DAF to assist DCSA's development of a process to identify and prioritize critical SLTT LEA organizations whose CHRI contributions are needed for vetting individuals as part of the BI process and who might benefit from federal assistance. The study was conducted through three tasks.

In Task 1, we reviewed relevant instructions, policies, directives, manuals, and guidance on past and current activities related to accessing SLTT CHRI for BIs. We also reviewed research on nationwide CHRI collection.

We obtained data from DCSA on current collection efforts by DCSA government and contract staff. To supplement the data acquired from DCSA and to validate the analysis of that data, we gathered detailed information on processes for information gathering, process impediments, and potential remedial actions via interviews with DCSA and SLTT experts. In this stage, we interviewed experts across DCSA in Federal Investigative Records Enterprise, Field Ops, LELO, and headquarters to gain insight on DCSA practices and challenges. We also interviewed subject matter experts at the Integrated Justice Information Systems (IJIS) Institute for insight on SLTT CHRI–sharing capability.

For Task 2, using the information collected in Task 1, we identified data elements that were required to support a DAF for DCSA to inform application of future federal assistance. We coordinated with the sponsor to gather and aggregate relevant data on both DCSA and SLTT CHRI collection and consistent information shortfalls. We cleaned and organized the data for further analysis, assessed the quality and usability of the data, and determined whether analysis from existing data provided reliable results. This task was critical because the identification and structuring of data elements was necessary to the data analysis and, more importantly, was the basis on which the DAF was built and will be maintained for continued use by DCSA.

Finally, in Task 3, we used the information gathered in earlier tasks as inputs and developed the DAF as an Excel-based tool. In developing this tool, we considered the goal of enabling DCSA leadership to make data-driven decisions regarding where improved cooperation and application of federal resources could mitigate CHRI collection issues and thereby improve the efficiency of the vetting process. In pursuit of this goal, we identified where impediments to cooperation presented the greatest problems for the vetting process and, through use of sophisticated analysis and visualization tools, identified specific areas where DCSA investment in equipment, systems, or training for SLTT organizations might ameliorate these issues and streamline access to historical CHRI data.

Assumptions and Limitations

This study has several noteworthy assumptions and limitations. First, although DCSA interacts with multiple levels of LE, published research and data are not generally available at the local level. Data are also often lacking for tribes and territories. The DAF created for this study uses data generally aggregated at the state level; Washington, D.C., is included when information is available. The DAF is intended to provide broad guidance rather than an in-depth understanding of how to obtain CHRI from local agencies. Second, data presented in this report and the DAF are a snapshot in time; data collection occurred between May 2020 and December 2021. When possible, data sets from before the coronavirus disease 2019 (COVID-19) pandemic were used to avoid bias because of pandemic policy changes or exemptions that do not reflect standard practice. Third, because of data set limitations (i.e., maintenance and data entry), this version of the DAF does not include detailed information of DCSA encounters with LEAs. Future versions of the DAF could integrate more DCSA-owned data to improve its utility in aiding DCSA decisions and improving value.

Structure of the Report

In this report, we summarize information and findings identified from our research, analysis, and evaluation and identify best practices and recommendations for supporting DCSA. In Chapter 2, we discuss findings from our review of CHRI practices across SLTTs. In Chapter 3, we discuss the data that comprise the DAF and how to use the tool. In Chapter 4, we provide a consolidated list of findings, recommendations, and conclusions for DCSA consideration. The DAF tool is provided as a separate Microsoft Excel file.

CHAPTER TWO

Varying SLTT CHRI Practices and Their Impacts on DCSA

DCSA routinely collaborates with thousands of LEAs across all states, tribes, and territories. These organizations, although bound by federal law, operate under diverse state laws, local policies, and even agency-specific practices. An awareness of broader trends in varying CHRI practices across SLTTs is needed so that DCSA can understand how best to support its diverse partners and ultimately achieve its aim of efficiently and effectively acquiring the information needed to conduct BIs. Whereas previous research efforts to assess extant CHRI practices across the United States focused mostly on SLTT-level performances and modernization efforts, this study focused on utilizing the SLTT organizations' CHRI performance data and analyzing them in a way that can provide valuable insights for DCSA concerning where investment in improved local processes can be of benefit.[1]

A comprehensive understanding of CHRI practices across the nation, particularly as they relate to individual SLTT LEAs and criminal history data repositories, can aid DCSA in identifying where the greatest hurdles exist for their investigators and how to be more agile in navigating those challenges. DCSA's BI process is not linear and only becomes more complicated if information is missing or inaccurate at different junctions of the process.[2] Given that every BI is unique, especially with CHRI data being pulled from different SLTTs, DCSA investigators and adjudicators cannot

---

[1] SEARCH, 2020a.

[2] See Bushway, Ligor, and Walsh, forthcoming, for a more detailed discussion on the DCSA BI process.

apply a one-size-fits-all approach for its processes. To ease the complication in this already nonlinear, multipart process, DCSA would benefit from federal and SLTT CHRI databases that are both complete and accurate, because they would reduce the barriers that an investigator or adjudicator might face during various steps of the investigation. Consequently, the efficiency and quality of the DCSA's criminal BI processes hinges on the quality and accessibility of CHRI data. In addition, the DCSA investigators interviewed for this report stated that they adopt different practices in interacting with different SLTTs, some of which can require extra time and effort depending on the locale. Therefore, it was important for this study to gather existing criminal justice policies (as they pertain to CHRI) from across the country to gauge the overall standards and quality of CHRI practices.

To do this, our literature review aimed to explore existing legislation and policies governing various aspects of CHRI across the SLTTs and aggregate this information into a single data set.[3]

Method

This study includes the results of publicly available reports pertaining to CHRI practices across the United States, particularly SEARCH reports, the IJIS Institute, and the DOJ.

The literature review targeted various aspects of existing LE practices, including legislation, regulations, policies, and practices as they pertain to CHRI across the SLTTs. For instance, LE laws such as cite and release, state definitions for felony and misdemeanor, fingerprinting requirements across the United States, and various other components were collected for the data set. We also reviewed CHRI practices to gauge overall CHRI data quality, including factors such as rap sheet quality, digitization of criminal records, database maintenance practices, and interstate record shareability, across the SLTTs.

[3] Most of the literature came from the DOJ, the IJIS Institute, and various SEARCH reports available to the public domain. See References for more information.

Findings by Section

Technological Systems and Web Portals

In general, the study found that many states were in the process of making major transformations in their current CHRI processes. Some of these critical changes included the updating of SLTT computerized criminal history (CCH) systems, automated fingerprint identification systems (AFISs), and state message switches.[4] These technological systems are critical for SLTTs to maintain accurate information for their respective CHRI repositories and to enable effective inter- and intrastate sharing of criminal information (see Box 1). Using the 2018 SEARCH findings, we found that the average age of CCH systems across the SLTTs was 7.23 years, message switch was 6.07 years, and AFIS 2.96 years.[5] Figure 2.1 shows the ages of each technology system.

In addition, having a robust and up-to-date technological system to track CHRI is particularly critical because of the increase in the number of individual records entered into the criminal history repository. From 2008 to 2018, the United States saw an increase in criminal history input by 21.9 percent.[6] Updated technological systems allow for greater accuracy and scalability when entering large amounts of data. Furthermore, thanks

BOX 1

Critical Indicator on Technological Systems and Web Portals

- Fees associated with criminal history reports for local or state criminal justice agencies

[4] The Bureau of Justice Statistics (BJS) routinely provides grant funding to SLTTs to update these systems through their National Criminal History Improvement Program and the NICS Act Records Improvement Program.

[5] Becki R. Goggins and Dennis A. DeBacco, *Survey of State Criminal History Information Systems, 2018*, Washington, D.C.: U.S. Department of Justice, No. 255651, November 5, 2020.

[6] Goggins and DeBacco, 2020.

FIGURE 2.1
SEARCH Survey Result of Technology Age Across SLTTs

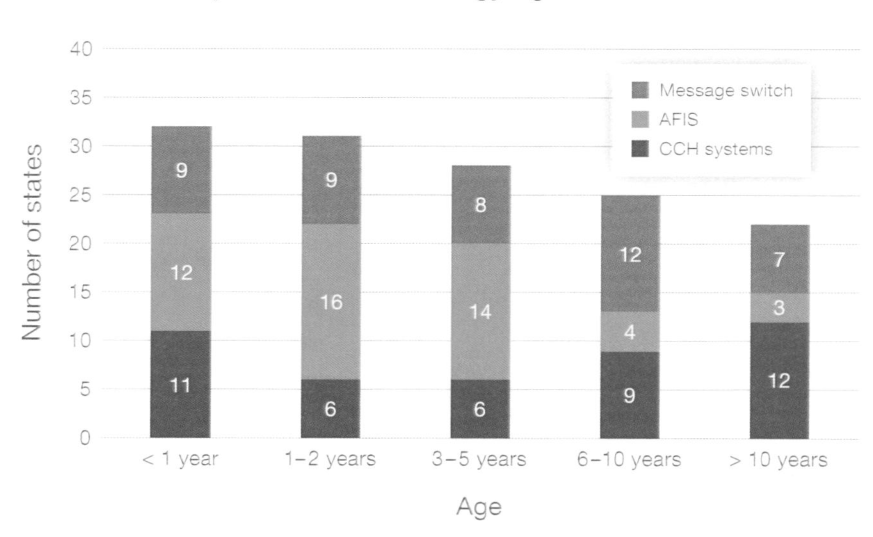

SOURCE: RAND analysis of Goggins and DeBacco, 2020.

to technological advancements allowing for cloud hosting and automation, CHRI data can now be more easily shared across systems, and inefficiencies caused by manual processes can be reduced with the latest technology, although regulatory barriers could inhibit these advancements. DCSA inherently benefits from SLTTs updating their technological systems because this eases CHRI data collection efforts.

Many local LEAs still maintain paper-based records, which means the only way to obtain their data is by conducting local field investigations. If adjudicators indicate missing data in the investigation that requires record checks at the local level, this could potentially cause additional administrative burden and delays for DCSA.

Furthermore, the policies for sharing these background checks and criminal history reports with investigators varied widely across SLTTs, mostly in terms of their accessibility and fees. Navigating the fee structure and access to SLTT-level CHRI data can be problematic for DCSA. In some instances, states do not offer web-based criminal justice background checks at all, and other states limit them from public usage.[7] In addition, fees for the web-based services varied, ranging from flat fees of $4 to $25 per check.[8] Some states also charged varying amounts depending on the type of web-based check required. SLTTs charge varying fees for criminal history reports requested by local and state criminal justice agencies.[9] Fees for access to the digital database vary widely among different SLTTs from free to thousands of dollars. The fee structures are also varied—for instance, some states charge for individual types of reports, some charge agencies annual or quarterly subscription-based access to their databases, and some charge per VPN access or even physical page count.

Laws on Expungement, Sealing, and Automatic Record Clearing

Policy and legislation governing CHRI policies across the SLTTs affect DCSA because different practices can complicate DCSA's ability to gather CHRI (see Box 2). Such gaps in CHRI data create increased work for investigators because they must contact local courts or use other methods of

BOX 2

Critical Indicators on Laws on Expungement, Sealing, and Automatic Record Clearing

- Authority to expunge, seal, or set aside convictions
- Automatic record clearing

[7] DCSA does not use the state web-based portals, but rather uses III, Nlets, and occasionally direct terminal access in situations where Nlets is not available. DCSA might consider completing a review of the utility of existing databases used for direct terminal access to identify best practices and determine which most effectively provide complete investigatory data. This insight can be used by DCSA to help direct funding to organizations so that any upgrades to digital capabilities synchronize with DCSA needs.

[8] Data were derived from Goggins and DeBacco, 2020. Fee structures could have changed since data were gathered for this report.

[9] Although Nlets and III do not require a fee, DCSA might need to pay fees in states in which it does not use Nlets.

investigation to fill in these gaps. Sometimes these gaps become permanent, affecting the overall quality of the investigation.

Different laws governing regulations for expungements and sealing for various SLTTs could create inconsistent reports for different BIs depending on the state. Although DCSA is entitled to sealed records under Title V, some LEAs do not know or recognize this and might not share them at the local law check stage.[10] For instance, Texas does not provide relief of felony convictions and authorizes sealing of misdemeanor cases only when the individual has no prior convictions or, for more serious misdemeanors, deferred dispositions after a two-year waiting period. In Missouri, the law allows expungement of all non–Class A felonies and all misdemeanors (subject to list of exceptions for more violent crimes).[11] In Alaska, there is no statute or regulation that outlines authority for its LEAs to seal or expunge adult conviction or nonconviction records.[12] Although DCSA should have access to expunged and sealed records, compliance issues could occur if LEAs do not understand DCSA's federally granted authorities in this space. Furthermore, SLTTs have varying levels of stringency governing their expungement and sealing policies (see Figure 2.2), which could potentially affect the type of records DCSA receives (or does not receive) for its BIs.

SLTTs also have a variety of laws governing the prevalence of automatic record clearing, which allows for records to be cleared without a petition (and often for free). More than 20 states allow some type of automatic record relief.[13] For example, in 2020, Michigan passed a law that allows for a variety of nonconvictions, misdemeanors, and some felonies to be automatically cleared from individuals' records. These records are expunged automati-

[10] DCSA has access to these sealed records through III and Nlets.

[11] Revised Statutes of Missouri, Title 39, Section 610.140, Expungement of Certain Criminal Records, Petition, Contents, Procedure—Effect of Expungement on Employer Inquiry—Lifetime Limits, August 28, 2018.

[12] Collateral Consequences Resource Center, "Restoration of Rights Project: State-Specific Guides to Restoration of Rights, Pardon, Expungement, Sealing & Certificates of Relief," webpage, undated.

[13] National Conference of State Legislatures, "Automatic Clearing of Records," webpage, July 19, 2021.

FIGURE 2.2
Types of Expungements and Sealing Practices Across States by Frequency

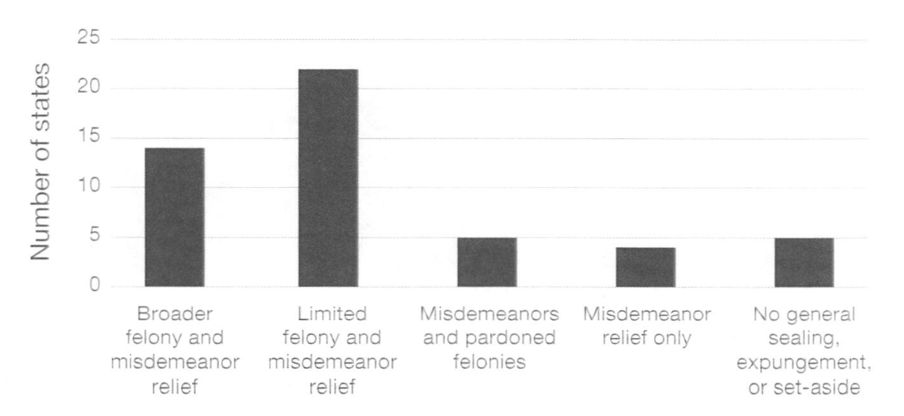

SOURCE: CCRC, undated.

cally without a petition after seven to ten years (depending on the seriousness of the conviction).[14]

Record-Keeping Practices and Data Quality Audits

Variance in record-keeping practices across SLTTs affects DCSA's investigation process in several ways and ultimately can have a negative effect on the adjudication process if these practices keep investigators from fully gathering all pertinent CHRI (see Box 3). State differences in record-keeping practices include the management of warrants, citations, and protection orders. These files, although not related to a formal conviction, provide DCSA with important context for their BIs. Only five states maintain statewide citation files that augment the state criminal history repository.[15]

[14] Restoration of Rights Project, "Michigan Restorations of Rights & Record Relief," webpage, Collateral Consequences Resource Center, December 3, 2021.

[15] Goggins and DeBacco, 2020.

BOX 3

Critical Indicators on Record-Keeping Practices and Data Quality Audits

- Warrant file maintained by locality
- Indictment information posted to the criminal history record by locality
- Statewide citation file maintained by locality
- Protection order file maintained by locality
- Protection orders entered into NCIC
- Internal audits of data quality

The primary purpose of statewide protection orders and citation files is not to provide data for DCSA's investigation; however, these files include information that would not be available in the state repository and thus help fill in gaps in CHRI.[16] If states do not maintain such files, DCSA can still receive this information, but it must be entered directly into NCIC by LEAs or courts. If the information is entered into local systems, electronic repositories then allow for quicker confirmation of protection orders and citations, and automation increases the efficiency of the transmission of data to NCIC (and reduces the likelihood of human error), helping to prevent delays in the investigation process.[17] Maintaining files for these types of data helps ensure that there are no gaps or inconsistencies in an individual's CHRI. Similarly, maintaining warrant and protection order files enables repositories to capture more data points and reduces gaps in BI information. Washington, D.C., and 40 states have statewide protection order files containing more than two million records.[18]

[16] Becki Goggins and Karen Lissy, "Survey Insights Blog Series #3: State Cite and Release Practices and Statewide Citation Files," SEARCH, December 30, 2020.

[17] Susan Keilitz, *Order Repositories, Web Portals, and Beyond: Technology Solutions to Increase Access and Enforcement*, Williamsburg, Va.: National Center for State Courts, 2020.

[18] Goggins and DeBacco, 2020.

Given the saliency of felonies in DCSA investigations, the ability of states to flag felonies is important because it lets investigators immediately recognize whether an offense is a felony in the state of conviction, again reducing delays in the investigation.[19] Forty-one states can immediately flag felonies to identify whether an individual was convicted of a felony.[20]

Data quality also has an important effect on states' ability to report CHRI, and it follows that states whose data are audited are likely to have better-quality data. The FBI audits states periodically to ensure the quality and accessibility of data. Repositories can contribute to these efforts through internal audits. Twenty-eight states perform internal audits of data quality, but only nine states and Washington, D.C., perform internal audits multiple times per year. Four states perform such audits annually, and the other 14 perform them only every two or more years.[21]

Cite and Release Practices

For misdemeanor offenses, most states' LE officers "ticket" the offender with a citation in lieu of a formal arrest; this is called *cite and release*. Cite and release programs provide various types of benefits that make the process more efficient, specifically reducing the time required for LE officers to complete an encounter. However, because citation does not typically warrant fingerprinting, the biometric identifier required to generate a criminal history arrest entry is not provided (see Box 4). As a result, the arrest records for offenders who were cited and released often goes missing. This is prob-

BOX 4

Critical Indicator on Cite and Release Practices

- Cite and release arrests without fingerprints

[19] Becki Goggins, *Findings and Emerging Trends from the 2016 Survey of State Criminal History Information Systems*, SEARCH, April 2, 2018.

[20] Goggins and DeBacco, 2020.

[21] Goggins and DeBacco, 2020.

lematic, because it exacerbates an existing issue of states being unable to link court dispositions to an arrest record.[22] Furthermore, studies have shown that cite and release arrestees are less likely to appear in court; failure-to-appear rates range from approximately 15 percent to 40 percent.[23] This can be especially problematic for DCSA adjudicators, because they check for both arrest records and final dispositions (if any exist) in the individual's investigation report.[24] Although DCSA does not count states as being non-compliant if the number of missing arrest records for less serious offenses remains low, accurate BIs require accurate arrest records.[25]

Furthermore, cite and release practices are inconsistently adopted across the SLTTs. Only half the states use cite and release for violations and misdemeanors; 16 states use them for misdemeanors and select felonies; and four states, Washington, D.C., and Guam do not routinely use cite and release practices. Furthermore, 54 percent of states have statutes that require courts to order the offender who was cited and released to be fingerprinted either before or after the initial court hearing, although four states have departmental rules in place to do the same. The remaining states do not have policies in place for fingerprinting cited and released offenders.[26] In total, 46 states cite and release in some fashion (see Figure 2.3).[27]

---

22 Mark Perbix, *Unintended Consequences of Cite and Release Policies*, SEARCH, June 2013.

23 Goggins and Lissy, 2020.

24 5 U.S.C. 9101.

25 SEARCH, 2020a.

26 Goggins and Lissy, 2020.

27 Goggins and DeBacco, 2020.

FIGURE 2.3
2018 SEARCH Cite and Release and Fingerprinting Data

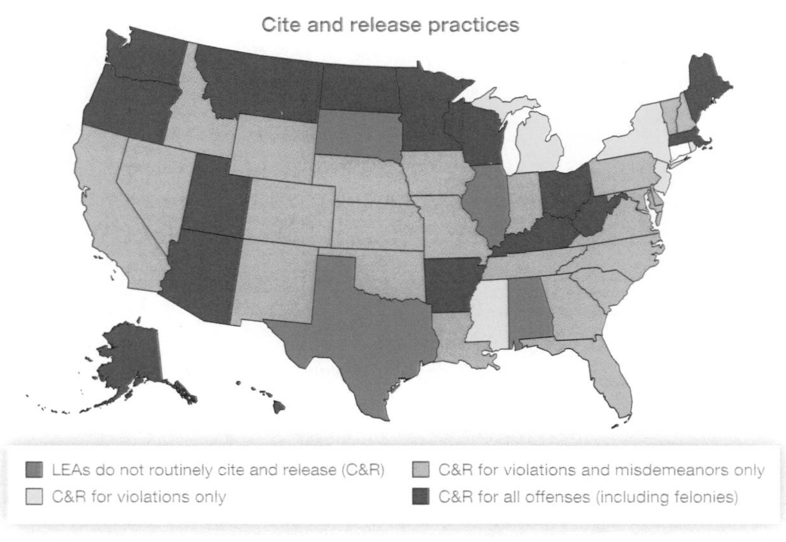

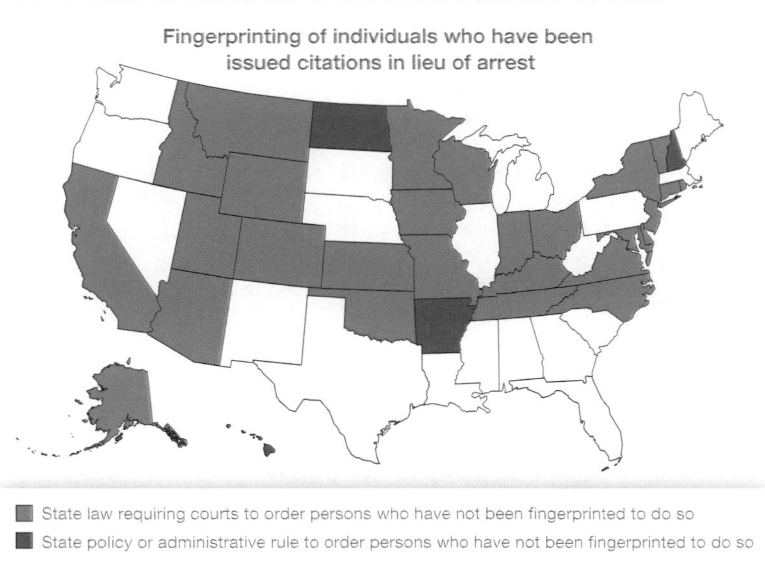

SOURCE: Goggins and Lissy, 2020.

States' Participation in Interstate and National Exchange of CHRI

DCSA adjudicators rely on Nlets and the III (including NCIC) to obtain CHRI for their BIs.[28] The FBI's III links to Next Generation Identification (NGI) and serves as the data-indexing (integration) mechanism for biometric and nonbiometric CHRI searches.[29] CHRI in the FBI III also receives Universal Control Numbers (UCNs) for records and tracking.

NCIC has shared data management responsibilities with federal, state, local, tribal, territorial (FSLTT) entities and links to the FBI's III. The NCIC contains 21 additional descriptors (e.g., domestic violence and fugitive status).[30] The FBI maintains the operational status of the NCIC (e.g., telecommunications infrastructure), whereas state-level criminal justice information system agencies are responsible for providing NCIC access to local LE organizations. State LE agencies are responsible for updating information in the NCIC or reporting errors to DCSA representatives.

[28] Sean D. Bushway, Ryan Andrew Brown, Dulani Woods, and Lee Remi, *Comparison of Criminal History and Private Consumer Reporting Agency Background Checks: Implications for National Security Clearance Adjudications*, Santa Monica, Calif.: RAND Corporation, 2022, Not available to the general public; RAND and Federal Bureau of Investigation, Criminal Justice Information Services Division, and Bureau of Justice Statistics, *Comparison of Criminal-History Information Systems in the United States and Other Countries*, Washington, D.C.: Bureau of Justice Statistics, No. 253816, April 2, 2020; William J. Krouse, *Gun Control: National Instant Criminal Background Check System (NICS) Operations and Related Legislation*, Washington, D.C.: Congressional Research Service, R45970, October 17, 2019; U.S. Code, Title 34, Section 40316, National Crime Prevention and Privacy Compact, October 9, 1998; Office of the Attorney General, *The Attorney General's Report on Criminal History Background Checks*, Washington, D.C.: U.S. Department of Justice, June 2006; and Peter M. Brien, *Improving Access to and Integrity of Criminal Records*, Washington, D.C.: U.S. Department of Justice, Bureau of Justice Statistics, No. 200581, July 2005.

[29] Unlike the NGI, data contained with the III includes "only names and personal identification information relating to persons who have been arrested or indicted for a serious criminal offense anywhere in the country" or nonbiometric data. See Code of Federal Regulations, Title 28, Part 20, Criminal Justice Information Systems, May 20, 1975.

[30] FBI, "Privacy Impact Assessment for the Fingerprint Identification Records System (FIRS) Integrated Automated Fingerprint Identification System (IAFIS) Outsourcing for Noncriminal Justice Purposes—Channeling," webpage, May 5, 2008.

Nlets can be used to search both state CHRI databases and state Department of Motor Vehicle records. Nlets is not used to search FBI records, but it does search the same databases as III (with name, not fingerprint).[31] Nlets receives state-level funding; each state has one Nlets point of contact and is governed by regionally appointed Nlets directors.[32] DCSA has validated use of Nlets for security clearance investigations in 25 states and employs traditional information gathering (e.g., investigator interviews) for the remaining states.[33]

SLTT LEAs' compliance and participation in national compacts that allow for ease of sharing CHRI data are critical to maintaining accurate information in these repositories (see Box 5).[34] As of July 2019, 34 states ratified the National Crime Prevention and Privacy Compact, and 10 states, American Samoa, Guam, and Puerto Rico have signed the memoranda of understanding (MOU).[35] Ratifying the compact effectively allows the FBI and ratifying states to participate in the noncriminal justice access program

BOX 5

Critical Indicators on States' Participation in Interstate and National Exchange of CHRI

- CHRI retrievable through III
- Nlets participant validated by DCSA

[31] Bushway et al., 2022.

[32] Harsin, Teri, "Nlets 101," blog, Nlets, September 30, 2021.

[33] DCSA, "DCSA NLETS Validation List 20210223.xlsx," data file, 2021.

[34] These compacts primarily concern state-to-state information-sharing. Although this does not directly concern DCSA, the broader information-sharing environment and states' willingness to enter such agreements could suggest increased willingness to cooperate with DCSA and other federal CHRI sharing initiatives.

[35] Becki Goggins, "Survey Insights Blog Series #7: States' Participation in the National Systems and Programs that Facilitate Interstate Exchange of Criminal History Records," SEARCH, April 19, 2021.

of III.[36] The compact creates a legal framework by which states can access III for noncriminal justice purposes because it supersedes any conflicting state laws for interstate exchange of CHRI.[37] It also allows states to help set III policy through the compact council. Furthermore, states must sign the compact to participate in the state-owned National Fingerprint File (NFF), partially because the compact reduces NFF's operation costs by eliminating duplicate maintenance of records by the FBI.[38] Signing an MOU allows states to declare support for the rules, policies, and procedures of the compact without undertaking the lengthier process of ratifying it. Like ratification, signing of an MOU enables states to support the III system purpose codes allowing the exchange of CHRI. However, states with MOUs are not allowed to participate in the council, and they might not fully implement the compact without ratification.[39] Figure 2.4 shows the current compact state and states that signed the MOU.

The FBI's NGI system shares data linkage with the III and appends existing FBI records based on state-owned information.[40] The NGI houses biometric CHRI (e.g., iris-scans, palm prints, and body tattoos) that is not included in other FBI or state-owned repositories. Each CHRI identity receives an FBI UCN to establish unique tracking identification. The NGI's National Rap Back service notifies other FSLTT agencies and departments when activities are reported to NGI on "persons who are licensed

[36] The FBI responds to noncriminal justice background check inquiries on behalf of the state if the state doesn't support the purposed code provided for the request.

[37] DCSA has access to these sealed records through III and Nlets.

[38] Although 34 states have ratified the compact, only 20 fully participate in the NFF. The states that do not fully participate rely on outdated AFISs, which sometimes lack the technical capacity to meet the NFF participation qualifications (established by the compact council), or rely on legacy AFISs that cannot meet fingerprint identification standards that enable full participation. The root cause for both reasons is that states cannot or will not complete expensive AFIS upgrades for access to the NFF alone but prefer to wait for their scheduled AFIS enhancements. Goggins, 2021; and National Crime Prevention and Privacy Compact, *Frequently Asked Questions Regarding the National Crime Prevention and Privacy Compact Act of 1998*, Version 5.1, May 2015.

[39] Goggins, 2021.

[40] The NGI is replacing legacy integrated automated fingerprint identification systems. See RAND, FBI, and BJS, 2020.

FIGURE 2.4
Compact and MOU Signatory States

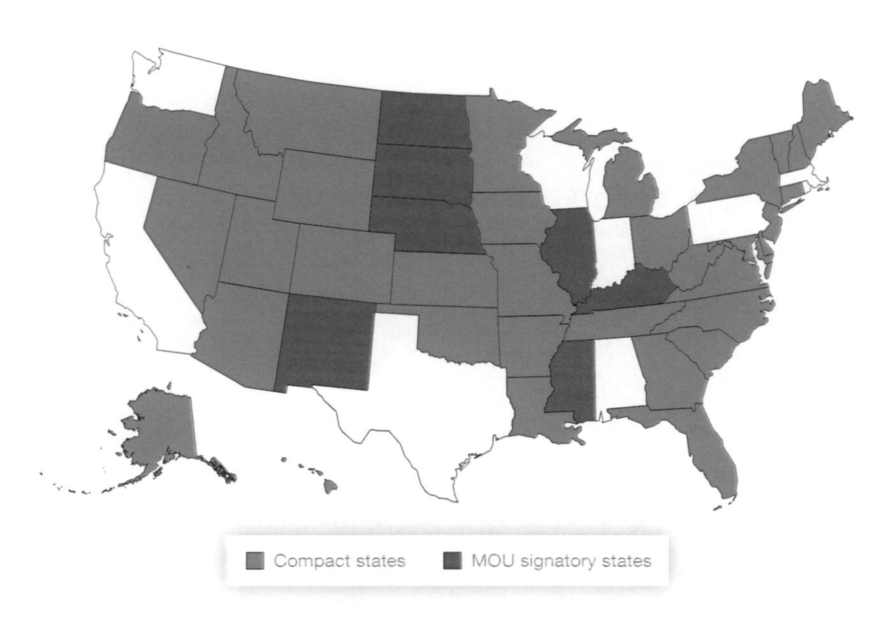

SOURCE: Goggins, 2021.

or employed (e.g., schoolteachers and day-care workers) or who are under criminal justice supervision or investigation, eliminating the need for repeated background checks on a person from the same agency."[41]

Disposition Reporting Practices

The literature review revealed significant variation in current CHRI practices across SLTTs in the United States when it comes to various collection of data and reporting processes. One such concern was the SLTT LEAs' interim disposition reporting practice, which varied widely across the United States (see Box 6). Interim disposition informs the status of an arrest or charges that have not reached its conclusion. Most outcomes result in adjudications

[41] FBI, *Law Enforcement Records Management Systems (RMSs) as They Pertain to FBI Programs and Systems*, Washington, D.C.: U.S. Department of Justice, undated; OAG, 2006; and RAND, FBI, and BJS, 2020.

BOX 6

Critical Indicators on Disposition Reporting Practices

- Repository receipt of final case dispositions from local prosecutors
- Receipt of court dispositions by automated means
- Charge-tracking information (interim dispositions) on the criminal history record to show case status through the criminal justice process collected by locality
- Attempts to locate missing disposition information before responding to fingerprint-based noncriminal justice inquiries

or dismissals, but during this process of prosecution and reaching a final verdict, interim dispositions help to inform courts and LEAs of the status of the case. Interim dispositions reporting is particularly helpful because final dispositions typically take significant time to process . SEARCH found that only 50 percent of states posted interim dispositions, whereas only about 33 percent of states and Guam incorporated their criminal case filings in their CHRI (see Table 2.1).[42]

Local prosecutors can provide final dispositions that otherwise might be overlooked. For example, prosecutors often dispose of cases by declining to file charges or through diversion programs.[43] When prosecutors decline to file charges, the charges are never brought before the court, meaning the prosecutor might be the sole source of information on the outcome of the disposition. This also happens when an individual completes a prefiling diversion program. As of 2018, there were 34 states in which local prosecutors sent final court dispositions to the state. Eight of those 34 states received those dispositions through an automated prosecutors' case management system (CMS) centralized at the state level.[44] Seven states received final disposition through prosecutors' CMS at the local level. There are also

---

[42] Becki Goggins, Karen Lissy, and Mark Perbix, "Survey Insights Blog Series #4: A Closer Look at Reporting Interim Dispositions," SEARCH, February 4, 2021.

[43] SEARCH, 2020a.

[44] Goggins and DeBacco, 2020.

TABLE 2.1
Interim Disposition and Indictment and Active Case Reporting

Practice[a]	States	No. of States
Interim dispositions (including felony case filings and indictments)	Arizona, Colorado, Delaware, Georgia, Hawaii, Maryland, Michigan, Minnesota, Mississippi, Missouri, Nevada, New Hampshire, and Oklahoma	13
Other interim dispositions (not including felony case filings and indictments)	Arkansas, Florida, Illinois, Kansas, Maine, Nebraska, New Jersey, New York, North Dakota, Rhode Island, South Dakota, Texas, Utah, Vermont, and Wisconsin	15
Only indictments are reported as interim dispositions	Alabama, Ohio, and South Carolina	3 and Guam
No interim dispositions	Alaska, California, Connecticut, Idaho, Indiana, Iowa, Kentucky, Louisiana, Massachusetts, Montana, New Mexico, North Carolina, Oregon, Pennsylvania, Tennessee, Virginia, Washington, West Virginia, and Wyoming	19 and Washington, D.C.

SOURCE: Goggins, Lissy, and Perbix, 2021.

NOTE: [a]Practice refers to the information that is reported to the state criminal history repository.

19 states in which local prosecutors send dispositions as paper files, and 11 states use a mix of automated and paper means. In addition, 28 states report on interim dispositions to update ongoing cases. Sixteen states add this information to the criminal history record. These indictment practices provide more information to DCSA, filling in gaps in CHRI.

Interim disposition reporting is important for DCSA because if records from central CHRI databases show incomplete or missing dispositions, the investigating officer needs to reach out individually to the courts to obtain information, which can create administrative burden. In addition, in 2018, 13 states reported that one-quarter or more of all dispositions received might not be linked to specific arrest records.[45] Missing or inaccurate dispositions and arrest records can stem from a variety of reasons, such as failure by the arresting agency to submit reports, errors in data submission,

[45] Goggins, 2018.

erroneous fingerprinting, and data entry backlogs. Data entry backlogs can be problematic: In 2018, 20 states reported more than two million court dispositions that were yet to be submitted in the repository. Automation can typically help with remedying these types of inefficiencies, and based on information from 2018, it appears that most states are adopting the new automation system to report court dispositions in their state repositories (see Figure 2.5).[46] These types of data gaps can create larger investigative burden and potentially yield inaccurate results for DCSA's adjudicators.

FIGURE 2.5
Automated Court Disposition Reporting, 2018

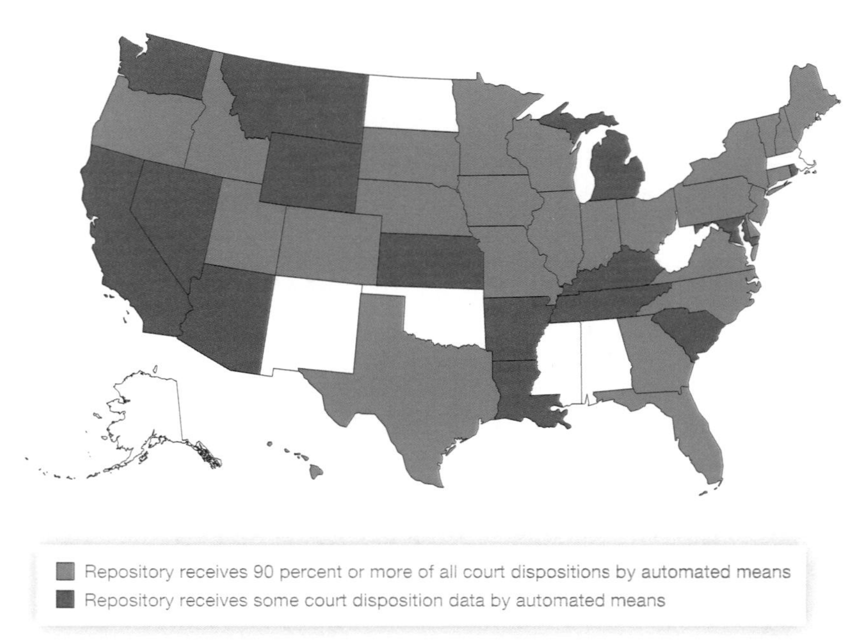

SOURCE: Goggins and Perbix, 2021.

[46] Becki Goggins, and Mark Perbix, "Survey Insights Blog Series #5: 2018 Survey Gauges Level of Disposition Reporting by Courts and Local Prosecutors," SEARCH, February 23, 2021.

Limitations

Despite our efforts to be exhaustive in the data collection, the aggregated data set had multiple components that could not be filled because of lack of data availability. This was particularly true for U.S. territories where data sets were often missing. It's critical to capture accurate CHRI data for U.S. territories, particularly for places with large military recruitment and presence such as Guam.[47] In addition, because the aggregated data set looked only at CHRI practices at the state level, it does not capture the more detailed practices at the county, city, or township levels, which affects DCSA's data collection practices.[48] Given that BIs for national security typically require more in-depth probing of past criminal history records of individuals, unveiling CHRI practices at the local level might provide DCSA with a better understanding of how to obtain CHRI data from local agencies.

For future DCSA studies, CHRI practices at the state, local, and territorial levels should be explored in-depth and should include criminal data repository systems, plans for updating existing systems, and collection of DCSA data to gauge how often local-level data are engaged in the overall BI process.

In addition, these data sets should be seen as valid information only as a snapshot in time. Most of the data used were based on SEARCH findings from 2018 that were officially published in 2020, then implemented into this report in 2021. Furthermore, the COVID-19 pandemic might also have affected how LEAs updated their technological platforms or carried out their policy revisions to update their CHRI practices.

How Will This Information Be Used for the DAF?

In our literature review, we summarized existing legislation and policies governing various aspects of CHRI across the SSLTs. In so doing, we identi-

[47] Jon Letman, "Guam: Where the US Military Is Revered and Reviled," *The Diplomat*, August 29, 2016.

[48] Analyzing CHRI data at the local township level was out of the scope for this research. In addition, compiled data at the lower local levels across all SLTTs were not available.

fied multiple critical indicators regarding SLTT CHRI collection and retention that appeared particularly relevant for DCSA's activities and for which we were able to find data for most, if not all, of the states. This data has been aggregated into a single repository, discussed in the next chapter.

CHAPTER THREE

Decision Aid Framework

This DAF identifies the activities that reduce the time and cost of accessing SLTT CHRI. The DAF's main objectives are to:

- Identify opportunities for DCSA resource investment to increase the efficiency and effectiveness of the background check vetting through greater access to CHRI.
- Identify localities that could leverage additional resources to improve provision of CHRI to DCSA to improve the efficiency and effectiveness of the background check vetting.[1]

In constructing the DAF, we drew from a broad suite of documentation, literature reviews, other data, and interviews to develop a multivariable ranking system that aims to support DCSA in achieving its decision objectives. The DAF ranks localities based on 24 indicators across six factors that influence how each locality actively participates in the background vetting process. The end ranking is derived from an indexed score that is weighted according to DCSA priorities. In this chapter, we describe the components and use of the DAF.

Terminology

To develop the DAF, we used a set of assumptions and terms to describe the core theoretical operations of the multivariable ranking system. In the

[1] We currently define *localities* as the 50 states and Washington, D.C., but other territorial, tribal, and local data could be added in future iterations of the DAF.

following section, we define and describe the following assumptions and terms:

- **Indicator:** a variable that is used to capture variation between localities. The indicators were selected because of their importance to CHRI collection.
- **Factor:** an assigned grouping of indicators.
- **Score:** a value that is assigned to each indicator based on data type. For qualitative indicators we used a five-point scale; for quantitative variables, we broke scores up into quintiles and maintained the Boolean data on a 0/1 basis.
- **Index:** a way to compile one score from a variety of data sources and types that represents the overall condition of a measured outcome. These composite measures allow for the interlinking of different data types in a summary of the overall "macro" condition of the topic or concept being indexed. An index is helpful for giving the decision maker a high-level view across factors in the index.
- **Weighting:** a system that changes the raw values of the index to account for preferences, priority, and asymmetry of information in the system. We used a standard 0–1 weighting scale for simplicity.

Method

Our DAF uses a multiparameter ranking method, which we have derived from multiattribute utility theory (MAUT).[2] MAUT is an analytic framework that was developed to evaluate individual decisions made by a single decisionmaker under risk given an incomplete set of information (asymmetrical information). For our purposes, we moved beyond an individual

[2] For more on MAUT, see Detlof Von Winterfeldt and Gregory W. Fischer, "Multi-Attribute Utility Theory: Models and Assessment Procedures," in Dirk Wendt and Charles Vlek, eds., *Utility, Probability, and Human Decision Making*, Dordrecht, the Netherlands: Springer, 1975, pp. 47–85. For an example of MAUT applied to performance management, see Alisha D. Youngblood and Terry R. Collins, "Addressing Balanced Scorecard Trade-off Issues Between Performance Metrics Using Multi-Attribute Utility Theory," *Engineering Management Journal*, Vol. 15, No.1, 2003, pp. 11–17.

decisionmaker and examined a larger set of parameters that could inform an institutional-level decision under asymmetrical information.[3]

Our study used the data sources and critical indicators identified in the literature review to build the DAF. To capture the level of data variation we were interested in, we combine two types of variables. The first type is an *indicator*—a discrete measurement of specific information available to a decisionmaker, which can be scored. The second type is a *factor*, which is a grouping of relevant indicators. These factors are then indexed to form a base score by which each locality can be ranked.

The multivariable ranking approach in the DAF combines both qualitative and quantitative information to help the decisionmaker to deploy resources in support of program decisions and to track progress across time. We used a three-step process to develop our DAF (see Figure 3.1). Each step in the process included a reevaluation of the data, literature, and documentation (described in Chapter 2), allowing for iteration and refinement. Not all data was complete for all 50 states and Washington, D.C.; indicators for which no relevant information was available received missing data scores.

- **Step 1:** Data was scoped, sorted, and organized into a system review structured by data type.
- **Step 2:** Indicators were scored and normalized across data type, and the localities were organized by factor scores composed of component indicator scores.

[3] Similar approaches have been used in other RAND studies examining decision support (sometimes referred to as *robust decision making*) for developing clinical specialties, identifying potential regional impacts of climate change, and planning for COVID-19 risk and vulnerabilities across major U.S. cities. See Cheryl L. Damberg, Justin W. Timbie, Douglas S. Bell, Liisa Hiatt, Amber Smith, and Eric C. Schneider, *Developing a Framework for Establishing Clinical Decision Support Meaningful Use Objectives for Clinical Specialties*, Santa Monica, Calif.: RAND Corporation, TR-1129-DHHS, 2012; Michelle E. Miro, *Identifying and Planning for Vulnerabilities in the San Bernardino Valley Municipal Water District's Water Management Plans*, Santa Monica, Calif.: RAND Corporation, TL-A1284-1, 2022; and Pedro Nascimento de Lima, Robert J. Lempert, Raffaele Vardavas, Lawrence Baker, Jeanne S. Ringel, Carolyn M. Rutter, Jonathan Ozik, and Nicholson Collier, *Reopening California: Seeking Robust, Non-Dominated COVID-19 Exit Strategies*, Santa Monica, Calif.: RAND Corporation, EP-68758, 2021.

FIGURE 3.1
Indicator and Factor Analytical Breakdown

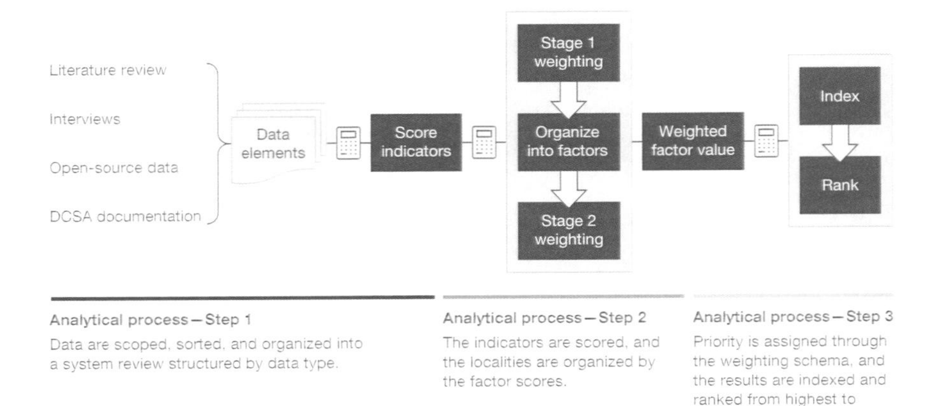
Literature review
Interviews
Open-source data
DCSA documentation
Data elements
Score indicators
Stage 1 weighting
Organize into factors
Stage 2 weighting
Weighted factor value
Index
Rank
Analytical process—Step 1
Data are scoped, sorted, and organized into a system review structured by data type.
Analytical process—Step 2
The indicators are scored, and the localities are organized by the factor scores.
Analytical process—Step 3
Priority is assigned through the weighting schema, and the results are indexed and ranked from highest to lowest weighted index score.

- **Step 3:** Priority was assigned through the weighting schema, and the results were indexed and ranked from highest to lowest weighted index score.

Indicators and Factors

Indicators are variables used to capture variation between localities in their ability to support CHRI collection. The indicators are drawn from literature reviews, source documents, and open-source data, as described in Chapter 2. After gathering and constructing the indicators discussed previously, we next worked to develop the factors. These factors group similar indicators together to draw more tangible insights and allow decisionmakers to make targeted judgments. For example, if a locality has a highly functioning administrative process system with limited administrative burden, then the total demand for DCSA resources should be lower than a locality with a higher level of administrative burden. Thus, a locality with factor value higher than the overall administrative burden average is in less need for resources and, therefore, ranks higher than a locality with a lower score and a higher assistance demand. Table 3.1 lists all the current factors and associated indicators.

TABLE 3.1
DAF Factor and Indicator Alignment

Factor	Indicators	Definition
Administrative burden	• Repository receipt of final case dispositions from local prosecutors • Rap sheet performance • Cite and release arrests without fingerprints • Authority to expunge, seal, or set aside convictions • Receipt of court dispositions by automated means	The degree to which requirements, information collection, and administrative rules limit or impede DCSA's criminal background check process
Systematized barriers	• Warrant file maintained by locality • Indictment information posted to the criminal history record by locality • Statewide citation file maintained by locality • Protection order file maintained by locality • Charge-tracking information (interim dispositions) on the criminal history record to show case status through the criminal justice process collected by locality	The systematic barriers that DCSA might face when attempting to report or collect relevant information as prescribed by statute
Legislative and policy complexity	• Fees associated with criminal history reports for local or state criminal justice agencies • Protection orders entered into NCIC • Use of Rapid ID • Internal audits of data quality • Attempts to locate missing disposition information before responding to fingerprint-based noncriminal justice inquiries • Felony flagging capability • Automatic record clearing	Complexities of the legislative policy environment that drive decisions and political and legislative choices by the localities that shape the type of reporting that can occur
Interaction with federal government	• CHRI retrievable through III • Nlets participant validated by DCSA	The extent to which a locality participates in federal programs that are designed to ease the background check process

Table 3.1—Continued

Structural conditions	• Rural population • Locality expenditures on police • Total crime	Structural factors that contribute to DCSA's process (quantitative values)
DCSA adjustment	• Law check quantity • Law check priority	Value that reflects a locality's relative impact on DCSA's CHRI collection efforts

SOURCE: RAND analysis of data sources listed in the DAF Data Sources section in References.

NOTE: For more detailed information on the indicators, please see the DAF.

The following sections describe the groupings of indicators under specific factors, how we evaluated the individual indicators, and how this information contributed to furthering DCSA's mission. The groupings of indicators were subjective and determined by our analysis of how those indicators would most affect DCSA.[4]

Administrative Burden

In interviews, DCSA headquarters and field staff emphasized the limited resources—regarding both LEA funding and personnel availability—to take on the additional tasks of CHRI information-sharing. Obtaining CHRI, particularly in complicated situations or where additional steps are required to gather information, can therefore constitute an administrative burden to LEA staff. For our administrative burden factor, the variables are designed to capture the degree to which requirements, information collection, and administrative rules either limit or impede DCSA's criminal background check process. The indicators used to measure administrative burden include rap sheet performance, whether the repository receives any final case dispositions from local prosecutors, whether the locality conducts cite and release arrests without fingerprints, locality expungement or sealing, and the percentage of court dispositions added by automatic means.

[4] Indicators can be moved from one factor category to another with minor adjustments to the Excel formulas in the DAF.

We considered rap sheet performance to contribute to administrative burden because it is an intricate part of the information collection process. Because DCSA needs these criminal background histories to conduct an adequate review, weaknesses in the rap sheet materially contribute to the amount of time the agency needs to spend on any given individual BI. Rap sheet performance was evaluated on how many weaknesses were found: the more weaknesses, the lower the score.[5]

- Localities with no weaknesses in formatting received a 5.
- Localities with a few weaknesses that should not materially affect the security clearance review received a 4.
- Localities with weaknesses that should cause moderate impairment received a 3.
- Localities with weaknesses that will cause significant impairment received a 2.
- Localities with weaknesses that will severely impair the process received a 1.

We coded whether the repository receives any final case dispositions from local prosecutors as a simple yes or no, with a binary (0 or 1) score. Localities in which the repository does receive the final case disposition received a 1, and those in which the repository does not receive the final case disposition received a 0. Similar to the scaled variables, this indicator shares the attribute of contributing to the information collection process for DCSA; therefore, we included these indicators as part of administrative burden. Any requirement or administrative rule that makes it easier for the agency to collect the necessary data contributes to this process.

We also coded the extent to which localities have the authority to expunge, seal, or set aside convictions. The more expunged or sealed records there are, the more opportunities there are for the relevant locality to refuse to report or to be unable to report information on the incident. Furthermore,

[5] *Weaknesses* are the number and severity of issues that could be improved in rap sheets. See SEARCH, *State Criminal History Records Profiles: Prepared for the Performance Accountability Council (PAC) Program Management Office (PMO)*, April 10, 2020b.

localities that make it easier to get expungements would be more likely to refuse or be unable to turn over expunged or sealed records. Accordingly, this indicator was included in administrative burden. Localities were given the following scores based on their level of authority:[6]

- Localities with broader felony and misdemeanor relief received a 1.
- Localities with limited felony and misdemeanor relief received a 2.
- Localities with misdemeanor and pardoned felonies relief received a 3.
- Localities with misdemeanor relief scored a 4.
- Localities with no general sealing or set aside scored a 5.

We identified which localities cite and release without fingerprinting. As these incidents create a gap in the individual's CHRI that investigators must use additional time and resources to fill, they fall within administrative burden. Localities were scored on a scale from 1 to 4 based on the offenses for which localities cite and release without fingerprints.[7]

- Localities that cite and release without fingerprinting for criminal offenses, including felonies, scored a 1.
- Localities that cite and release for violations and misdemeanors scored a 2.
- Localities that cite and release only for violations received a 3.
- Localities that do not cite and release received a 4.

We also examined the method by which court dispositions were added to a digital repository. When courts use automatic means to report dispositions to repository, DCSA theoretically can access that repository to gather all needed information without having to spend time with individual clerks and courts. Accordingly, this indicator was included under administrative

[6] Restoration of Rights Project, "50-State Comparison: Expungement, Sealing, & Other Record Relief," webpage, Collateral Consequences Resource Center, last updated March 2022.

[7] Goggins and DeBacco, 2020.

burden. This indicator includes data on the percentage of court dispositions that were reported to locality repositories through automated means.[8]

- Localities with .00 through .20 received a 1.
- Localities with .21 through .40 received a 2.
- Localities with .41 through .60 received a 3.
- Localities with .61 through .80 received a 4.
- Localities with .81 through 1 received a 5.

Systematized Barriers

Outdated and outmoded reporting systems are some of the frequent problems DCSA confronts when aggregating the information necessary to conduct a review. Localities vary in the degree to which they deploy updated technology and whether they follow reporting standards. The indicators in this factor are designed to assess the barriers to entry that DCSA might face when attempting to report or collect relevant information. Several indicators comprise this factor: whether the locality maintains a warrant file, collects charge-tracking information on the criminal history record to show case statuses through the criminal justice process, posts indictment information to the criminal history record, maintains a statewide citation file, and maintains a protection order file. All indicators above were scored through a simple binary (yes or no) coding scheme. Localities that collect or post the information to the criminal history record received a yes (coded as a 1), and the localities that do not maintain such information received a no (coded as a 0).

The five indicators under systematized barriers reflect the degree to which DCSA will need to rely on manual transfers of data and, therefore, reflect the amount of additional work DCSA will need to do because the information is not centrally located or is incomplete. If the locality does not maintain a warrant file or post indictment information to the criminal history record, DCSA would have to find a different way to aggregate that data. This is especially important for localities that DCSA has flagged for in-person visits because of a history of missed felony convictions on criminal history records.

---

[8] Goggins and DeBacco, 2020.

Legislative and Policy Complexity of Criminal Reporting Systems

Legislation and other policy created by the states and local LEAs can also shape the criminal reporting environment. For example, some localities charge federal agencies for the use of data streams. This factor comprises the following seven indicators:

1. Fees associated with criminal history reports for local or state criminal justice agencies
2. Whether the locality is currently employing Rapid ID
3. Whether the locality has felony flagging capability
4. Whether protection orders are entered into the NCIC
5. Whether internal local audits of data quality are contributing to audits the locality conducted
6. Whether localities attempt to locate missing disposition information from repositories before responding to fingerprint-based non-criminal justice inquiries
7. Whether state or local laws allow for automatic record clearing.

Fees associated with the criminal history reports use a scaled score because of the wide disparity in amounts that localities charge. The premise behind the coding was: the higher the fee, the more of an obstacle for DCSA to obtain the record, and the lower the score.

- Localities without a charge received a 5.
- Localities with potential charges or local LE caveats received a 4.
- Localities with low recurring charges, guaranteed charges, or multiple conditions received a 3.
- Localities with recurring charges in the several hundreds of dollars range received a 2.
- Localities with fees that could go up to several thousand received a 1.

We employed the yes or no classification scheme (with a corresponding binary [0 or 1] coding) for the remaining indicators.

Interaction with Federal Government

State and local governments interact with the federal government for a variety of reasons related to CHRI sharing. This factor is designed to measure the extent to which the localities participate in federal programs that are designed to ease the background check process. Because of this, we measure the factor with two yes or no indicators: whether CHRI data can be retrieved through III and whether DCSA has validated the use of Nlets for the locality. We considered these elements to be a part of the interaction between levels of government because III participation and Nlets validation allow DCSA rapid and efficient access to CHRI.

Structural Conditions

Interviews with DCSA administrative and field personnel revealed several structural conditions that affect the background check process. We captured these structural factors with several quantitative elements: the percentage of the population that lives in a rural area, the total number of crimes committed in the state, and how much the locality spends on police.

Interviews revealed that a locality's urban/rural divide affects the security clearance process. Interviewees discussed how operating in rural areas can cause bottlenecks because of where the records were stored. One respondent said that they went to a rural LE office only to be told that the records were stored at the courthouse. Such instances increase the amount of time DCSA needs to spend to conduct any given search, both because of travel and the need to interact with more than one office (explaining what DCSA is doing, why the records are necessary, etc.). All data on the urban/rural divide comes from the 2010 U.S. Census (the 2020 version was unavailable when we compiled the data).[9] The specific measure we took from the U.S. Census was the percentage of the population that lives in a rural area as a proxy for the number of people DCSA might have to investigate in any given area. The most rural localities received a 1, the least rural localities received a 5; we then normalized these scores to fall between 0 and 1. The coding scheme is based on DCSA feedback in interviews: The localities that are the

[9] U.S. Census Bureau, "Percent Urban and Rural in 2010 by State," data file, undated.

most rural raise the likelihood of DCSA having to interact with the small offices, with diverse record keeping, and with additional travel time.

The next structural indicator we included was the total number of crimes committed in each locality. The total number of crimes in an area could affect the amount of time and effort DCSA spends compiling the background check portfolio. We constructed this indicator by combining the total number of property crimes with the total number of violent crimes.[10] We then took the average number of total crimes per locality from 1960 to 2012 to arrive at a uniform measure. Localities with the highest crime totals received a 1, and the localities with the lowest crime totals received a 5; all scores were then normalized to 0–1. We employed the average number of crimes in each locality to account for the likely difficulties DCSA could face when obtaining and evaluating arrest records and criminal histories. When a locality has had a significant number of crimes over a sustained period, DCSA will need to both locate those records and ensure their accuracy. As the number of records DCSA has to review increases, there is a greater likelihood that files will be missed or the records obtained will not be accurate. With a large number of crimes, there is also an increased possibility of obtaining false records. For example, if DCSA is reviewing the security clearance packet for someone named John Smith in California, there are likely to be far more arrest records under that name compared with an investigation of John Smith in Wyoming.

The last structural indicator we included was locality expenditures on police.[11] Localities that spend a larger percentage of funds on the police might have more developed technological infrastructures to handle and process DCSA's requests more efficiently. Following the established procedure, we took the average expenditures on police per locality and separated into quintiles. The localities with the highest expenditures received a 5, and the localities with the lowest expenditures received a 1; scores were then normalized. We similarly employed average year numbers to account for trickle down effects, and legacy problems and to prevent outliers. For

[10] Whitcomb, Ryan, Joung Min Choi, and Bo Guan, "State Crime CSV File: From the CORGIS Dataset Project," data file, October 19, 2021.

[11] Data can be downloaded from Tax Policy Center, "State and Local General Expenditures, Percentage Distribution," webpage, August 27, 2021.

example, if Kentucky spent little money on its police from 2010 to 2020, but then boosted the budget in 2021, the average year figure would account for the problems still present in 2021. Investments in technological infrastructure and administrative personnel take time to effect lasting change. The 2021 budget would not necessarily have produced tangible improvements that would need to overcome the problems introduced by a decade of neglect, especially compared with localities that spent an average higher amount from 2010 to 2020 but did not spend as much in 2021. Therefore, we employed the average locality expenditures on police to reflect as well as possible the locality's investments in its LE capabilities.

Index Generation

To arrive at a final rank for each locality, indicators first had to be transformed into a format that was cross-comparable and able to be aggregated into factors. Indicators described above that were not scored on a 0–1 scale were normalized to values between 0 and 1. These normalized scores were then multiplied by weights (based on user-set priorities). The raw factor score is generated by averaging the final normalized and weighted component indicator scores. The values for all the normalized factor scores were then divided by 5 (the number of factors described above) and multiplied by 100 to arrive at the index score.

DCSA Adjustment

The index score was further modified before arriving at the final rank. We incorporated DCSA data to assess both the overall quantity and the priority of investigations that DCSA conducts in each locality. This factor does not directly assess a locality's ability to share CHRI and so was not included in the index above. Rather, these indicators were used to adjust the index score to emphasize appropriately a locality's importance to DCSA's investigation process.

Using the total law check counts in each locality between January 1, 2020, and February 28, 2020, as a proxy for investigation quantity, we calculated the percentage of each locality's investigations out of all investigations.[12]

[12] *Law checks*—investigator-led searches conducted by field operations staff—were used as a proxy for overall investigations due to data availability. Law checks are initi-

As a proxy for investigation priority, we calculated the percentage of high-priority (Tier 5 and Tier 4) law checks for each locality between January 1, 2020, and February 28, 2020. Calculated in this way, the higher values indicate that the locality conducted more Tier 5 and Tier 4 investigations and is therefore more important to DCSA's investigation process. As other metrics are calculated so that *low* scores indicate that more assistance is needed, the scores for this factor were then subtracted from 1 so that lower values of the investigation quantity and priority proxy metrics would likewise indicate the locality is more important to the investigation process. The indicator values for this factor were then averaged and multiplied by the index. The result is that localities that were more important to DCSA's investigations would have a *lower* final adjusted index score, whereas localities that were less important to DCSA's investigations would increase their rank and receive a *higher* final adjusted index score. The final adjusted index scores highlight localities that could benefit from increased DCSA assistance *and* whose resulting improvement could have a higher impact on DCSA's CHRI collection. The values of this score were ranked, providing the final overall rank for the localities.

ated only when LEAs do not respond to mailed CHRI inquiry forms for investigations that are tier 2 or higher, when information obtained from preliminary checks indicates that further investigation is needed, or when the applicant self-reported a criminal history record that was not obtained through preliminary checks or the mailed inquiry form. Therefore, the values in this data set are biased toward LEAs that have not provided CHRI through the preferred (digital or mail) routes, so the data set emphasizes the LEAs that would benefit from increased DCSA assistance. The law check data files from January 1, 2020, to February 28, 2020, were the last file of law checks that occurred before the COVID-19 pandemic. See DCSA, "Law Checks 2020.01.01 - 2020.02.28.xlsx," data file on law checks conducted by the DCSA between January 1, 2020, and February 28, 2020, 2020a.

Using the DAF

Overview

The DAF was developed in Excel. This approach enables DCSA to leverage a well-known software to interpret and unpack the data analysis. The DAF has ten interworking tabs that allow the user to access essential information. These ten tabs are:

- **Banner:** This tab states the title and authorship information.
- **License Agreement:** This tab describes terms of use.
- **Home:** This tab reproduces the information in this section to introduce the user to the tool and its purpose of aiding DCSA in making institutional priority– and data-driven decisions.
- **Scorecard:** The scorecard is the final view of the ranked localities based on the scored factors, ordered from most to least likely to benefit from DCSA investment. In addition, scores are broken out by factor to allow the user to identify which elements drive the overall ranking score.
- **Factor Settings:** DCSA can assign weights based on institutional priorities to the various factors accounted for in the tool. Each factor is weighed on a scale of 0 to 1, and the total sum of all weights must equal 1. This allows DCSA to customize the priority assigned to each of the different factor categories we identified, which can alter the overall rankings of the localities.
- **Indicator Settings:** This tab allows the user to assign weights based on individual indicator priorities.
- **Codebook:** The codebook helps to unpack the different data types included in the tool. Each indicator is defined by type, long name, short name, and a description of how it was calculated based on the raw data.
- **Calculations:** This tab provides the critical initial data used and includes the calculation and treatment described above in each subcolumn.
- **Data Sources:** This tab provides the references for the data used in the tool.
- **Change Log:** This tab notes changes made to the tool.

One of the primary goals in developing this tool is to make it accessible and relevant to the work done at DCSA. The Excel-based DAF produces a scored ranking system that can help DCSA analyze and synthesize insights from their data sets to improve resource allocation and, ultimately, the overall clearance process by identifying localities that might need additional support.

Visualizing the Data

To visualize the outcomes of our ranking system, we used a scorecard that reports the weighted factor score across each factor per locality, the DCSA adjustment value, the adjusted index score, and its associated comparative rank. The weighted factor score is the raw factor score multiplied by the weight assigned in the factor settings tab. We then take these weighted factor scores, index them, adjust the index value according to the DCSA adjustment score, and finally rank each locality from a high score (i.e., the locality has a lower need for additional resources) to a low score (i.e., the locality has a higher need for additional resources).

Examining one row in the ranking system, we can see shading variance across cells. A darker colored cell indicates that the value of the locality exceeds the average value of that specific factor, and a lighter cell indicates that the value of the locality is lower than the average value of the specific factor. For example, if a state has an administrative burden score of 0.20 (as in Figure 3.2), and the average administrative burden score for the data set is 0.19, the state's cell would be dark gray because it exceeds the threshold of 0.19. Conversely, if that state's administrative burden score was 0.18, the cell would then turn light gray because it would be below the threshold of 0.19.

The scorecard has nine columns, including the column listing the localities (see Figure 3.3). Then there are the identified factor columns, the score for the DCSA adjustment value, the adjusted index value column, and finally a ranking column. Next to the matrix is a scale interpretation that shows that a lower rank and lighter color means fewer resources are needed and a higher rank and darker color indicates more resources are needed. Please note that the rank system includes data only for localities that we have in the database. Localities can be tied for the same rank.

FIGURE 3.2

Weighted Factor Score, Indexing, and Ranking

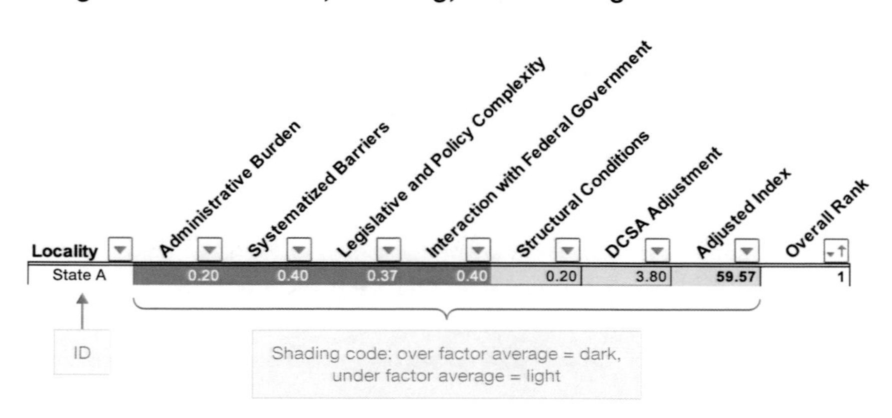

Locality	Administrative Burden	Systematized Barriers	Legislative and Policy Complexity	Interaction with Federal Government	Structural Conditions	DCSA Adjustment	Adjusted Index	Overall Rank
State A	0.20	0.40	0.37	0.40	0.20	3.80	59.57	1

FIGURE 3.3

Overview of the Scorecard

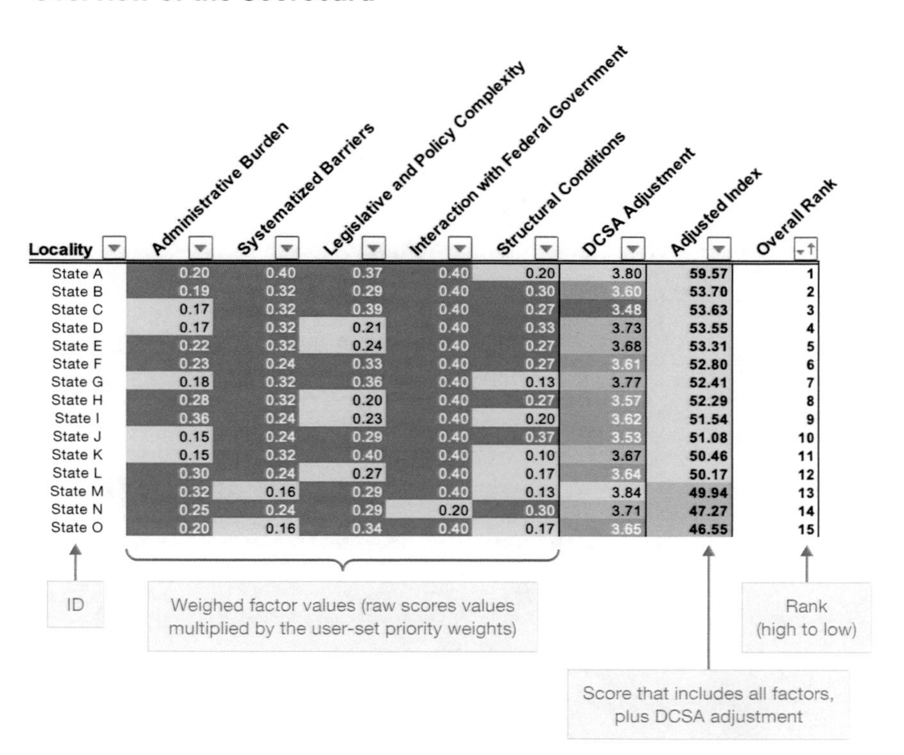

Locality	Administrative Burden	Systematized Barriers	Legislative and Policy Complexity	Interaction with Federal Government	Structural Conditions	DCSA Adjustment	Adjusted Index	Overall Rank
State A	0.20	0.40	0.37	0.40	0.20	3.80	59.57	1
State B	0.19	0.32	0.29	0.40	0.30	3.60	53.70	2
State C	0.17	0.32	0.39	0.40	0.27	3.48	53.63	3
State D	0.17	0.32	0.21	0.40	0.33	3.73	53.55	4
State E	0.22	0.32	0.24	0.40	0.27	3.68	53.31	5
State F	0.23	0.24	0.33	0.40	0.27	3.61	52.80	6
State G	0.18	0.32	0.36	0.40	0.13	3.77	52.41	7
State H	0.28	0.32	0.20	0.40	0.27	3.57	52.29	8
State I	0.36	0.24	0.23	0.40	0.20	3.62	51.54	9
State J	0.15	0.24	0.29	0.40	0.37	3.53	51.08	10
State K	0.15	0.32	0.40	0.40	0.10	3.67	50.46	11
State L	0.30	0.24	0.27	0.40	0.17	3.64	50.17	12
State M	0.32	0.16	0.29	0.40	0.13	3.84	49.94	13
State N	0.25	0.24	0.29	0.20	0.30	3.71	47.27	14
State O	0.20	0.16	0.34	0.40	0.17	3.65	46.55	15

Each of the five factors in Figure 3.4 is assigned a priority by the user in the third column; the DCSA adjustment has a separate weighting table that is independent of the factor weight settings. The first column shows the full name and abbreviation of each factor. The second column consists of the indicators that are included in the factor. The third column allows users to select the priority they want to assign the factor, which is displayed as a percentage in the final column.

Figure 3.5 shows a cluster of calculations for a single indicator, *Rural* (the short name for Rural Population), in the Calculations tab of the tool. The first column in the cluster is the data entry column, colored yellow in Figure 3.5. The next column is the normalized Rural score, labeled as *Ruraln*. The final column is the weighted and normalized score (according to the setting chosen in the Indicator Setting tab), labeled in Figure 3.5 as *RuralnW*. Each cluster is collapsible to the final normalized and weighted score value. The Calculations tab is intended to allow for specific changes

FIGURE 3.4

Overview of the Factor Weighting Schema

Factor Weight Settings

Name	Indicators	Priority	Normalized Weight
Administrative Burden (AB)	Repository receives final case dispositions from local prosecutors Rap sheet performance Cite and release arrests without fingerprints Has authority for expunging, sealing, or setting aside convictions Receipt of court dispositions by automated means	Medium	20%
Systematized Barriers (SB)	Locality maintains a warrant file Locality collects charge-tracking information (interim dispositions) on the criminal history record to show case status through the criminal justice process Locality posts indictment information to the criminal history record Locality maintains a statewide citation file Locality maintains a protection order file	Medium	20%
Legislative and Policy Complexity (LP)	Fees associated with criminal history reports for local/state criminal justice agencies Protection orders (PO) entered into NCIC Currently employing Rapid ID Repository conducts internal audits of data quality Repository attempts to locate missing disposition information before responding to fingerprint-based noncriminal justice inquiries Felony flagging capability Automatic record clearing	Medium	20%
Interaction with Federal Government (GI)	CHRI retrievable through the Interstate Identification Index (III) Nlets participant (validated by DCSA)	Medium	20%
Structural Conditions (SC)	Rural Population Locality Expenditure on Police Total Crime	Medium	20%

DCSA Adjustment Weight Setting

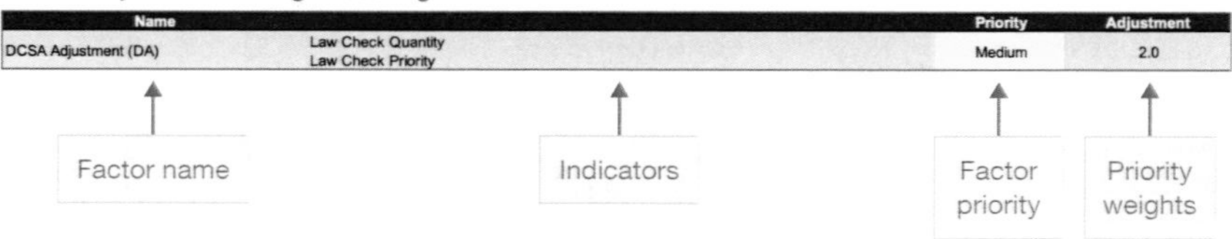

Name		Priority	Adjustment
DCSA Adjustment (DA)	Law Check Quantity Law Check Priority	Medium	2.0

NOTE: Yellow cells can be modified by users.

FIGURE 3.5
Overview of the Raw Score Elements

Locality	LocalityAB	Rural	Ruraln	RuralnW
State A	A	1	0.00	0.00
State B	B	2	0.25	0.50
State C	C	5	1.00	2.00
State D	D	1	0.00	0.00
State E	E	5	1.00	2.00
State F	F	4	0.75	1.50
State G	G	4	0.75	1.50
State H	H	4	0.75	1.50
State I	I	5	1.00	2.00
State J	J	5	1.00	2.00
State K	K	3	0.50	1.00
State L	L	5	1.00	2.00
State M	M	3	0.50	1.00
State N	N	5	1.00	2.00
State O	O	3	0.50	1.00
Name	ID	Raw score	Normalized score	Intrafactor weighted score

NOTE: Yellow cells can be modified by users.

and updates to the underlying data set and to facilitate transparency with our analytic approach.

Limitations and Future Work

Although we incorporated the most recent data available during the study period, a basic limitation of this study is that some data points might have become outdated as localities evolve their CHRI policies and more data becomes available. During the study period, some DCSA data could not be incorporated into the tool because of poor data quality. Given the nature of DCSA's work, other data could not be included in this iteration of the tool because they are protected. However, future versions of the tool could include more sensitive information and a greater degree of aggregated data collection. There are likely greater opportunities to integrate additional data into the tool as DCSA capabilities and needs evolve.

Another major limitation of the DAF is that the data could be included only at the state level, as data were not publicly available or consolidated for the 18,000 LEAs across the country. Data are also often lacking for tribes and territories, including both Guam and Washington, D.C., which are highly relevant for DCSA BIs. Because of this limitation in local, tribal, and territorial (LTT) data availability, the DAF is intended to provide broad guidance rather than an in-depth understanding of how to improve the process of obtaining CHRI from local agencies.

CHAPTER FOUR

Findings, Recommendations, and Future Research

Findings and Recommendations

- DCSA's CHRI collection is complicated by the myriad SLTT laws and policies regarding CHRI collection, maintenance, and sharing. In particular, different jurisdictions have different laws for expunging and sealing records, reporting arrests and dispositions, collecting fingerprints, and automatically clearing records.
  - Recommendations:
    - Improve data collection and enable cooperation by providing LEAs with access to federal funding and assistance that could be used to modernize criminal history data collection, management, and sharing capabilities. New data collection acquisition and management capability would directly support DCSA efforts to streamline, improve, and possibly accelerate personnel vetting processing.
    - Educate DCSA investigators on local CHRI laws and policies so they can address specific SLTT concerns with sharing CHRI.

- **SLTT information exchange between SLTT LEAs and DCSA is critical to DCSA's vetting process; unfortunately, many SLTT LEAs lack the personnel, knowledge, technology, or funding to enable compliance with their obligation to report CHRI in support of federal vetting.**[1] Outdated and outmoded reporting systems frequently slow or

[1] See 5 U.S.C. 9101.

obstruct DCSA's CHRI collection, and manual CHRI sharing could impose additional burdens on both LEA employees' and DCSA investigators' time.
  - Recommendations:
    - Improve cooperation by providing LEAs with access to federal funding and assistance that could be used to modernize criminal history data collection, management, and sharing capabilities.
    - Provide training, education, and direct assistance to SLTT LEAs for the purpose of streamlining access to historical and future CHRI data.[2]

- **The CHRI collection challenges that SLTTs face might be a product of SLTT laws and policies, the administrative burden the employees face in sharing CHRI, systemic barriers, government interaction through participation in federal programs, and structural conditions.** By focusing on the potential causes of insufficient CHRI sharing, DCSA would be better positioned to mitigate those specific collection issues. Additional information regarding the interactions of SLTTs with DCSA investigators (both staff and contractors) was not available to the researchers but would almost certainly improve the utility of the DAF.
  - Recommendations:
    - Use the DAF to drive decisions on where intensified cooperation and application of federal resources could best mitigate CHRI collection issues and ultimately improve the efficiency and effectiveness of DCSA CHRI collection.
    - Collect data on SLTT-DCSA interactions for all DCSA investigators and update the data when challenges are resolved or new information becomes available. Ensure data quality and accuracy by implementing standards for entry.

[2] For more detailed recommendations on SLTT LEA education and training, see Ligor et al., 2022.

Future Research

In addition to the issues noted in the findings, recommendations, and conclusions noted above, there are additional areas in which we suggest further research is warranted and that would likely yield valuable results. For future DCSA studies, CHRI practices at the state, local, and territorial levels should be explored in-depth and should include criminal data repository systems, plans for updating existing systems, and collection of DCSA data to gauge how often local-level data are engaged in the overall BI process. Awareness of the LTT practices will enable DCSA to fine-tune its ability to collect CHRI.

Abbreviations

AFIS	automated fingerprint identification system
BI	background investigation
BJS	Bureau of Justice Statistics
CCH	computerized criminal history
CHRI	criminal history record information
CMS	case management system
COVID-19	coronavirus disease 2019
DAF	decision aid framework
DCSA	Defense Counterintelligence and Security Agency
DoD	U.S. Department of Defense
DOJ	U.S. Department of Justice
FBI	Federal Bureau of Investigation
FSLTT	federal, state, local, tribal, territorial
III	Interstate Identification Index
IJIS	Integrated Justice Information Systems
LE	law enforcement
LEA	law enforcement agency
LELO	Law Enforcement Liaison Office (DCSA)
LTT	local, tribal, and territorial
MAUT	multiattribute utility theory

MOU	memorandum of understanding
NCIC	National Crime Information Center
NFF	National Fingerprint File (state-owned)
NGI	Next Generation Identification
Nlets	National Law Enforcement Telecommunications System
SEARCH	National Consortium for Justice Information and Statistics
SLTT	state, local, tribal, and territorial
UCN	Universal Control Number (FBI)

References

Brien, Peter M., *Improving Access to and Integrity of Criminal Records*, Washington, D.C.: U.S. Department of Justice, Bureau of Justice Statistics, No. 200581, July 2005.

Bushway, Shawn D., Ryan Andrew Brown, Dulani Woods, and Lee Remi, *Comparison of Criminal History and Private Consumer Reporting Agency Background Checks: Implications for National Security Clearance Adjudications*, Santa Monica, Calif.: RAND Corporation, 2022, Not available to the general public.

Bushway, Shawn, Douglas Ligor, and Stephanie Walsh, *Process Review: Overview, Initial Observations & Potential Responses for DCSA's State and Local CHRI Acquisition Process*, Santa Monica, Calif.: RAND Corporation, forthcoming.

CCRC—*See* Collateral Consequences Resource Center.

Code of Federal Regulations, Title 28, Part 20, Criminal Justice Information Systems, May 20, 1975.

Collateral Consequences Resource Center, "Restoration of Rights Project: State-Specific Guides to Restoration of Rights, Pardon, Expungement, Sealing & Certificates of Relief," webpage, undated. As of April 15, 2022: https://ccresourcecenter.org/restoration-2/

Damberg, Cheryl L., Justin W. Timbie, Douglas S. Bell, Liisa Hiatt, Amber Smith, and Eric C. Schneider, *Developing a Framework for Establishing Clinical Decision Support Meaningful Use Objectives for Clinical Specialties*, Santa Monica, Calif.: RAND Corporation, TR-1129-DHHS, 2012. As of July 22, 2022:
https://www.rand.org/pubs/technical_reports/TR1129.html

DCSA—*See* Defense Counterintelligence and Security Agency.

Defense Counterintelligence and Security Agency, "Law Checks 2020.01.01 - 2020.02.28.xlsx," data file on law checks conducted by DCSA between January 1, 2020, and February 28, 2020, 2020a.

———, *Report on the Defense Counterintelligence and Security Agency's Personnel Vetting Mission: In Response to H.R. 2500, the U.S. House of Representatives-Passed Fiscal Year 2020 NDAA*, June 2020b.

———, *DCSA Background Investigation Overview*, September 22, 2020c.

———, "DCSA NLETS Validation List 20210223.xlsx," data file, 2021.

Executive Order 13869, *Transferring Responsibility for Background Investigations to the Department of Defense*, Washington, D.C.: The White House, April 24, 2019.

FBI—*See* Federal Bureau of Investigation.

Federal Bureau of Investigation, *Law Enforcement Records Management Systems (RMSs) as They Pertain to FBI Programs and Systems*, Washington, D.C.: U.S. Department of Justice, undated.

———, "Privacy Impact Assessment for the Fingerprint Identification Records System (FIRS) Integrated Automated Fingerprint Identification System (IAFIS) Outsourcing for Noncriminal Justice Purposes—Channeling," webpage, May 5, 2008. As of April 15, 2022:
https://www.fbi.gov/services/information-management/foipa/privacy-impact-assessments/firs-iafis

Goggins, Becki, *Findings and Emerging Trends from the 2016 Survey of State Criminal History Information Systems*, SEARCH, April 2, 2018.

———, "Survey Insights Blog Series #7: States' Participation in the National Systems and Programs that Facilitate Interstate Exchange of Criminal History Records," SEARCH, April 19, 2021.

Goggins, Becki R., and Dennis A. DeBacco, *Survey of State Criminal History Information Systems, 2018*, Washington, D.C.: U.S. Department of Justice, No. 255651, November 5, 2020.

Goggins, Becki, and Karen Lissy, "Survey Insights Blog Series #3: State Cite and Release Practices and Statewide Citation Files," SEARCH, December 30, 2020.

Goggins, Becki, Karen Lissy, and Mark Perbix, "Survey Insights Blog Series #4: A Closer Look at Reporting Interim Dispositions," SEARCH, February 4, 2021.

Goggins, Becki, and Mark Perbix, "Survey Insights Blog Series #5: 2018 Survey Gauges Level of Disposition Reporting by Courts and Local Prosecutors," SEARCH, February 23, 2021.

Harsin, Teri, "Nlets 101," blog, Nlets, September 30, 2021. As of April 15, 2022:
https://nlets.org/resources/blog/nlets-101

Homeland Security Presidential Directive 12, *Policy for a Common Identification Standard for Federal Employees and Contractors*, Washington, D.C.: U.S. Department of Homeland Security, August 12, 2004.

HSPD—*See* Homeland Security Presidential Directive.

Keilitz, Susan, *Protection Order Repositories, Web Portals, and Beyond: Technology Solutions to Increase Access and Enforcement*, Williamsburg, Va.: National Center for State Courts, 2020.

Krouse, William J., *Gun Control: National Instant Criminal Background Check System (NICS) Operations and Related Legislation*, Washington, D.C.: Congressional Research Service, R45970, October 17, 2019.

Letman, Jon, "Guam: Where the US Military Is Revered and Reviled," *The Diplomat*, August 29, 2016.

Ligor, Douglas C., Shawn Bushway, Maria McCollester, Richard H. Donohue, Devon Hill, Marylou Gilbert, Heather Gomez-Bendaña, Daniel Kim, Annie Brothers, Melissa Baumann, Barbara Bicksler, Rick Penn-Kraus, and Stephanie Walsh, *Criminal History Record Information Sharing with the Defense Counterintelligence and Security Agency: Education and Training Materials for State, Local, Tribal, and Territorial Partners*, Santa Monica, Calif.: RAND Corporation, RR-A846-1, 2022. As of April 12, 2023: https://www.rand.org/pubs/research_reports/RRA846-1.html.

Miro, Michelle E., *Identifying and Planning for Vulnerabilities in the San Bernardino Valley Municipal Water District's Water Management Plans*, Santa Monica, Calif.: RAND Corporation, TL-A1284-1, 2022. As of July 22, 2022: https://www.rand.org/pubs/tools/TLA1284-1.html

Nascimento de Lima, Pedro, Robert J. Lempert, Raffaele Vardavas, Lawrence Baker, Jeanne S. Ringel, Carolyn M. Rutter, Jonathan Ozik, and Nicholson Collier, *Reopening California: Seeking Robust, Non-Dominated COVID-19 Exit Strategies*, Santa Monica, Calif.: RAND Corporation, EP-68758, 2021. As of July 22, 2022:
https://www.rand.org/pubs/external_publications/EP68758.html

National Conference of State Legislatures, "Automatic Clearing of Records," webpage, July 19, 2021. As of April 15, 2022:
https://www.ncsl.org/research/civil-and-criminal-justice/automatic-clearing-of-records.aspx

National Consortium for Justice Information and Statistics, *Challenges and Promising Practices for State Criminal History Repositories: Report to the Performance Accountability Council (PAC) Program Management Office (PMO)*, March 24, 2020a.

———, *State Criminal History Records Profiles: Prepared for the Performance Accountability Council (PAC) Program Management Office (PMO)*, April 10, 2020b.

National Crime Prevention and Privacy Compact, *Frequently Asked Questions Regarding the National Crime Prevention and Privacy Compact Act of 1998*, Version 5.1, May 2015. As of April 15, 2022:
https://ucr.fbi.gov/cc/library/compact-frequently-asked-questions

NCSL—*See* National Conference of State Legislatures.

OAG—*See* Office of the Attorney General.

Office of the Attorney General, *The Attorney General's Report on Criminal History Background Checks*, Washington, D.C.: U.S. Department of Justice, June 2006.

Perbix, Mark, *Unintended Consequences of Cite and Release Policies*, SEARCH, June 2013.

President's Task Force on 21st Century Policing, *Final Report of the President's Task Force on 21st Century Policing*, Washington, D.C.: U.S. Department of Justice, No. 248928, May 2015.

Public Law 116-92, National Defense Authorization Act for Fiscal Year 2020, December 20, 2019.

RAND Corporation and Federal Bureau of Investigation, Criminal Justice Information Services Division, and Bureau of Justice Statistics, *Comparison of Criminal-History Information Systems in the United States and Other Countries*, Washington, D.C.: Bureau of Justice Statistics, No. 253816, April 2, 2020. As of August 1, 2022:
https://www.rand.org/pubs/external_publications/EP68728.html

Restoration of Rights Project, "Michigan Restorations of Rights & Record Relief," webpage, Collateral Consequences Resource Center, December 3, 2021. As of April 15, 2022:
https://ccresourcecenter.org/state-restoration-profiles/michigan-restoration-of-rights-pardon-expungement-sealing/

———, "50-State Comparison: Expungement, Sealing, & Other Record Relief," webpage, Collateral Consequences Resource Center, last updated March 2022. As of April 15, 2022:
https://ccresourcecenter.org/state-restoration-profiles/50-state-comparisonjudicial-expungement-sealing-and-set-aside-2/

Revised Statutes of Missouri, Title 39, Section 610.140, Expungement of Certain Criminal Records, Petition, Contents, Procedure—Effect of Expungement on Employer Inquiry—Lifetime Limits, August 28, 2018.

RRP—*See* Restoration of Rights Project.

SEARCH—*See* National Consortium for Justice Information and Statistics.

Tax Policy Center, "State and Local General Expenditures, Percentage Distribution," webpage, August 27, 2021. As of April 15, 2022:
https://www.taxpolicycenter.org/statistics/state-and-local-general-expenditures-percentage-distribution

U.S. Census Bureau, "Percent Urban and Rural in 2010 by State," data file, undated. As of April 15, 2022:
https://www2.census.gov/geo/docs/reference/ua/PctUrbanRural_State.xls

U.S. Code, Title 5, Section 9101, Access to Criminal History Records for National Security and Other Purposes, 2011.

———, Title 34, Section 40316, National Crime Prevention and Privacy Compact, October 9, 1998.

Von Winterfeldt, Detlof, and Gregory W. Fischer, "Multi-Attribute Utility Theory: Models and Assessment Procedures," in Dirk Wendt and Charles Vlek, eds., *Utility, Probability, and Human Decision Making*, Dordrecht, the Netherlands: Springer, 1975, pp. 47–85.

Whitcomb, Ryan, Joung Min Choi, and Bo Guan, "State Crime CSV File: From the CORGIS Dataset Project," data file, October 19, 2021. As of April 15, 2022:
https://corgis-edu.github.io/corgis/csv/state_crime/

Youngblood, Alisha D., and Terry R. Collins, "Addressing Balanced Scorecard Trade-off Issues Between Performance Metrics Using Multi-Attribute Utility Theory," *Engineering Management Journal*, Vol. 15, No.1, 2003, pp. 11–17.

DAF Data Sources

Collateral Consequences Resource Center, "Restoration of Rights Project: State-Specific Guides to Restoration of Rights, Pardon, Expungement, Sealing & Certificates of Relief," webpage, undated. As of April 15, 2022:
https://ccresourcecenter.org/restoration-2/

Defense Counterintelligence Security Agency, "Law Checks 2020.01.01 - 2020.02.28.xlsx," data file on law checks conducted by the DCSA between January 1, 2020, and February 28, 2020, 2020.

———, "DCSA NLETS Validation List 20210223.xlsx," data file, 2021.

Goggins, Becki R., and Dennis A. DeBacco, *Survey of State Criminal History Information Systems, 2018*, Washington, D.C.: U.S. Department of Justice, No. 255651, November 5, 2020. As of April 15, 2022:
https://www.ojp.gov/pdffiles1/bjs/grants/255651.pdf

Greenspan, Owen, and Richard Schauffler, *State Progress in Record Reporting for Firearm-Related Background Checks: Fingerprint Processing Advances Improve Background Checks*, SEARCH and the National Center for State Courts, No. 250275, September 2016. As of March 25, 2022:
https://www.ojp.gov/pdffiles1/bjs/grants/250275.pdf

SEARCH, *State Criminal History Records Profiles: Prepared for the Performance Accountability Council (PAC) Program Management Office (PMO)*, April 10, 2020.

Tax Policy Center, "State and Local General Expenditures, Percentage Distribution," webpage, August 27, 2021. As of April 15, 2022:
https://www.taxpolicycenter.org/statistics/state-and-local-general-expenditures-percentage-distribution

U.S. Census Bureau, "Percent Urban and Rural in 2010 by State," data file, undated. As of April 15, 2022:
https://www2.census.gov/geo/docs/reference/ua/PctUrbanRural_State.xls

Whitcomb, Ryan, Joung Min Choi, and Bo Guan, "State Crime CSV File: From the CORGIS Dataset Project," data file, October 19, 2021. As of April 15, 2022:
https://corgis-edu.github.io/corgis/csv/state_crime/